Solving Division by Zero

Introducing the Unified Relativistic Number Theory

James Alexander
todayamerican@gmail.com

Created: March, 2020
Last updated: Feb 5, 2022

Preface

It may be hard to understand a lot of the material in this paper. The introduction is not going to make it any easier because I talk about my background, just so you know that this is not just some nonsense written by someone who never went to college. There are a lot of books written by people who don't even understand science. This is not one of them. If it were, there would not be so much math in this paper.

So, I wanted to mention the one fundamental truth which I realized after writing this book. The crux of why my theory makes sense.

I think if I am to claim a major leap in our understanding of math and physics, like Newton or Einstein, I need to also have found a profoundly different way of looking at the universe just like Copernicus, Galileo, Newton or Einstein. It is only by looking at the mundane from a different perspective that new insights can be made.

We have already made major strides in science in the 20th century. I think we have gone as far as we could go with the basic scientific principles, we have found out so far. One of the most important principles we have realized is that the Earth is not the center of the universe, and our planet is just another planet in a solar system on the periphery of a galaxy somewhere in a very large universe. The Earth is not the center of the universe.

Human senses have limits. We can only see a part of the electromagnetic spectrum. We can only hear from 20 Hz to 20,000 Hz. And there are limits to how much we can experience in the universe. Yet, our science has subtle, implicit biases which assume that what we can experience are absolutes. "Unless I see it, it cannot exist", "Unless I can experience it in some way with one of my senses, it cannot exist", etc.

We know about electromagnetic radiation which we cannot see and sound waves which we cannot hear. Science does consider these kinds of scenarios in many cases. But I have discovered some cases in math & physics which we have overlooked because they are so fundamental, mundane and foundational to science, that nobody has apparently realized that this kind of bias has stunted our understanding, limiting our scientific progress as a result.

This causes math to define absolutes like Zero and Infinity, in human terms. Just because something is very small to us, it is almost zero in size, or just because a galaxy is so large, its size is a very large number.

The problem with this kind of thinking in science is that, it is yet another way of thinking that humans with their limited sensory perceptions are at the "center" of the universe. Because we cannot think beyond this, we have hit many limits in science.

If anything, this paper proves that there is nothing special about humans or their senses. Our truths are not "absolute" truths. The concepts of zero or infinity, or very small and very large are not absolute, but "relative" to what humans think of as very small or very large.

Mathematics defined a number line starting from zero and ending in Infinity, at a time when humans did not even partially understand the concept or zero (nothing) or infinity (a very large number which we cannot quantify).

As science progressed over the past hundred years, we started using more and more sophisticated techniques to peer into the sub atomic world. The deeper we looked, the more we found and the search continues for the elusive fundamental building block of nature. Today with an electron microscope, it is possible to take a photo of individual atoms.

Where "nothing" was thought to exist before, today we know about the existence of atoms, electrons and protons.

How can we say that we understand the concept of "nothing", when we have not yet been able to find the fundamental building blocks or nature? There is something we can't yet find, underneath the fabric of reality - that is what physics tells us.

We know of the existence of black holes which are so dense that it contains more mass than the largest stars in existence. We cannot even imagine the total size of our planet, never mind the sun, or the myriad of super stars larger than the sun. These are entities which exist in the realm of "infinity" in terms of numbers which are so large as to be hard to understand or reach.

We know that the laws of nature do not seem to work within the event horizon of these black holes. A black hole is the nearest physical entity we have to something which is "infinite". Considering we cannot even make sense of a physical entity which has more mass than anything we can possibly imagine; how can we say that we understand the concept of infinity?

Mathematics defined concepts like zero and infinity, and failed to properly evaluate or understand equations or their corollaries which involve concepts which go beyond what humans could intuitively understand at the time. This is why they seem like meaningless equations - not because they are meaningless, but because we did not fully understand these concepts when we defined them, and started using them in mathematics.

When we use concepts, we do not fully understand in science, they may in certain scenarios produce results, we do not fully understand or may seem as nonsense. This does not mean that they are nonsense, it could mean that we do not understand them because of the flaws in our understanding of these concepts, and as you will see in this book, it may also mean that, the equations do make sense, when we look at them from a different perspective, which is beyond that of human intuition - because we see missing variables which can fill in the blanks, to make nonsense make sense.

We took concepts which were well beyond our ability to understand at the time (and still don't - as this book will explain) and placed them on either side of the number line. We then defined our number system somewhere in between. Then, we used them concepts in an equation, and arrived at some corollaries - why would it make any sense to us? - why would it be easy to understand?

Eyed through the lens of a purely scientific, unbiased approach, things which made no sense before, now starts to make perfect sense. What seemed to be nonsense before, now is perfectly sensible and logical.

Once you understand the concepts in this book, you will never look at science the same way again - you will always ask yourself "where else have we made the same fundamental mistakes?", "how much will need to be rewritten?", "where else have we left equations as undefined because we could not understand what they meant?"

For the first time ever, explained in excruciating detail, with innumerable, painstaking references from many, many obscure physicists and mathematicians who glimpsed at portions of the truth without quite getting there, this book will change the way you look at mathematics & physics, forever.

Contents

Solving Division by Zero 1

Preface 2

    Cover Image 6

Introduction 7

    Reviewing Scientific Papers 101: A Reminder to the new Generation 12

Conclusions 13

    The Unified Relativistic Number Theory: Why any number is literally the result of $\circledast \times \propto$ 13

        Naming the Theory 22

        Integrating Observers & Frames into the (new) Numeral System 23

        The Observer 23

        Relativistic Limits in Numbers 23

        Principle of Relativistic Number Atomicity 25

        Frame of Reference (Frame) 25

        Approaching Relativistic Limits: Why numbers trans mutate within equations 26

        Relativistic Isolation Theory 28

        On the matter of negative numbers 30

    Effects of Observers & Frames on Physics 30

        Time Dilation in adjacent frames 30

        Black Holes as Crossover Points to adjacent frame 30

        The behavior of Light in a Black Hole & The Relativistic Isolation Theory 31

        Multi-frame Theory & Black Hole Singularities 31

        Intuitionist Mathematics 31

        A Unified Field Theory 32

    The Proof is in the Pudding 34

    Quantum Number Theory/ Rule of Indeterminate Values 35

    Particle-Wave Number Theory/ Duality Principle of Indeterminate Numbers 37

        The meaning of dividing 1 orange by 0 39

    Where do the current Math Theories & Numeral System sit within the new System? 39

        On the matter of the Field Axioms 40

    Unified Relativistic Numeral System (URNS) 44

        The concept of 0 47

        Behavior of Atomic Zero $\odot$ $o1, p1$ 48

- **Transmutation of ⊙ $o1, p1$ into Indeterminate Zero ⊛ $o1, p1$** 49
- **The concept of ∞** 49
- **Introducing the concept of ∀ (Any)** 50

The Path to the Conclusions 53
- **Foreword** 53
- **Introduction** 54
- **Is the current Numeral System Infallible?** 56
  - **Analyzing 1/0 = ∞ without bias from Prior Art** 57
  - **What does this mean?** 58
  - **What is the wrong with this?** 58
  - **An Alternate View** 59
- **Prior Art** 60
- **New Math inspired from Physics** 61
- **Confronting the Paradox with the URNS** 61
  - **Meaning of ∨ = ⊛ x ∞** 63
  - **Why is zero in-determinate?** 64
  - **Summary** 65
  - **The question of ⊙/ ⊙** 66
- **Multi-dimensional Number Theory** 67
  - **References** 69
  - **Notes** 70

Cover Image

Ariane 5 / ATV-4 "Albert Einstein" - by the DLR German Aerospace Center

https://www.flickr.com/photos/48213136@N06/8967685954

This work is licensed under the Creative Commons Attribution 4.0 International License. To view a copy of this license, visit http://creativecommons.org/licenses/by/4.0/ or send a letter to Creative Commons, PO Box 1866, Mountain View, CA 94042, USA.

Introduction

I really did not want to put in an "origin story" as some people have called it within this paper, because it creates a bias. However, the Internet is full of scams and hackers cheating other people. And people don't know what they can trust or not. So, I realized that I need to give some context about myself before you read this book because after taking extensive feedback from people online, it is surprisingly clear that many concepts I have taken for granted that everybody knows (because Einstein and Stephen Hawking have been around for a while now) – they do not understand.

If you are not familiar with the basics of Relativity, Quantum Mechanics, Black Holes, Singularities, Hawking Radiation and the bending of light when it nears stars, you are not going to understand why many of the concepts in this book are true. So, I suggest that you spend some time reading the basics of these concepts if you want to truly understand this book. It will definitely help you understand what is going on.

I promise to spend more time putting in even more detail for people who are unable to understand the paper as-is. More links, more references, and I can see now, I need a lot of diagrams as well almost everywhere, so you can visualize what I am talking about. It is crystal clear to me, but looks like the message is not getting out to a western audience.

Ever since childhood, I have been interested in Relativity, Quantum Mechanics and Einstein. I remember sitting on the edge of a playground at school in 7th grade with smaller kids around me, talking about Relativity, Gravity & the bending of light near stars. I have been reading the Feynman Lectures in Physics since 7th Grade. It is possible, even though I personally think I don't know too much about physics (the actual mathematical proofs behind relativity, etc), I actually do know the concepts so well, for so long, that mathematicians can find it very hard to follow this paper, without being equally familiar in these areas themselves.

In 10th grade, I became the only student who has ever gotten 100% marks in Chemistry from one of our teachers. Students sat in the class where he mentioned what should be in the answers and gave the marks. Mine added to 100%, so he said: "Then, his answers must be perfect. Let us see if that is true". He went over every answer in front of the whole class and found no mistakes. I think I retain that record till date.

I drew the entire structure of DNA from memory in 12th grade in front of my whole class. My 12th grade project for CBSE (Central Exams in India) was a 300-page study on the psychology of students in my school taken with pain staking surveys from 6th to 12th grade (not an easy thing to do), and work done almost for the entire year. That project is not a book in my school library, and important reference book for all new teachers who join the school (it is a public, boarding school).

In Engineering College, my professor was open mouthed when I spent several days explaining all the equations and techniques for Nuclear Fusion for my seminar. Mine was the only seminar which went into this detail explaining exactly how the equations for Fusion worked and the challenges in getting it to work in a controlled environment.

So, I am not a high school pass out, or college failed guy without a job who has a lot of time on my hands to spend on a paper like this. I don't have that much time. And I have done advanced math all the way through final year of Engineering.

I don't think of myself as *intelligent*. In fact, I am just a normal guy, who was a very normal kid growing up, who got normal grades in most subjects except the ones he was interested in which included physics, genetics, Chemistry (except Organic Chemistry) & Math. Specifically for Math, I really got better at it since 9th grade, gaining interest as we learnt more advanced concepts like Calculus, and by the time I finished Engineering, my Dad was very surprised to see me get high marks in Engineering Math, when I used to struggle with the (still horrible) interest rates and such in 6th grade.

I was hard working though, and I could focus very well, and get very good marks in subjects I was interested in. I am a highly visual thinker and that has helped me over time (not so much as when I was a kid), to do better and better work as I grew up as well as in my career. If anything, I hope this book can ignite young minds to think beyond the box. There truly is no such thing as "impossible", because if you live with a problem and never give up, you will eventually find a solution to any and all problems.

I have a history of using available tools to unusual effect to solve hard problems. I once visited a factory where a machine was overheating, and I (not very great at Electronics) was able to utilize an existing IC in an unusual way to solve that problem. Later an Engineer working in that company who did his Masters came around to ask me: "How the heck did you figure that out?, I've never seen anyone use that IC in that manner".

I work in an exciting start-up which is very successful. I have encountered difficult and "impossible" problems many times over the past twenty years. I have a track record of solving unsolved problems. I studied math all the way through four years of my graduate course in Electronics Engineering. I am the guy who tends to live with a problem for years, trying over & over again until I get things to work, where everyone has failed before. This is exciting for me, and hence it is a passion of mine.

The only reason I have spent so much time on this paper is because so far I can't see any issues with any of the things I have written here. I can't find a major issue with any assumption. I can't find a major loophole in any theory. So, I am going to forge ahead no matter what people say. I don't care. I have proven myself over and over so many times – let the dogs bark till their throats go hoarse with it. I simply don't care.

I had the seed of an idea on Saturday, wrote up a preliminary draft in the next few days. Solicited a lot of comments online, modified the paper a lot. Through this entire process, this paper has only gotten stronger. I continue to work on it only because I see no mistake or issues with any of the concepts or theories mentioned here.

The paper maybe hard to understand, the language may not be very clear, but the concepts are strong. And I can explain and defend it to people who actually understand what is written here to begin with. If you don't understand what is written here to begin with, you got to ask me, before you jump to conclusions.

How did I start this paper? And why do I so vehemently argue some of the points? Here is the story behind it. Ramanujan has been quoted to have said that zero divided by zero maybe anything.

Ref: "National Mathematics Year: A Tribute to Srinivasa Ramanujan", from: https://bit.ly/2TOP8Iz

According to the famous English mathematician G.H. Hardy, Ramanujan was a natural genius like Euler and Gauss. Another British mathematician Littlewood commented, "I can believe that he is at least a Jacobi".

As Ramanujan said this statement, I have always believed that there must be something to it. This has forced me to revisit the problem of division by zero many, many times over the years. He famously produced equations with no proofs, which have been proven by others to be correct later.

I have failed to understand division by zero more times than I can care to remember. Twenty years later, I think I have it finally. It has taken a long time, and a lot of analysis to create this paper.

Indians invented zero, they also invented the current numeral system (yes, this is true – and simply a fact, don't get too antsy over a fact). It is only appropriate that an Indian take the current numeral system forward, after many, many years of stagnancy, and bring it into the 20th century of Einstein and Hawking.

This paper has been published because I want to get more critiques and commentary. Every comment, however acerbic, has helped improve this paper and get it to this point. I only ask that your commentary be very specific and to the point, so it is actionable – and either proves me wrong, or helps me improve the paper by incorporating the answer to your questions within it.

Needless to say, if I can't come up with a solution to one of the "issues" you find with this paper, it will be abandoned, if proven to be wrong. But vague opinions, and sentences are not anti-proofs.

This document is not (yet) a paper. It is deliberately written in lucid form so that it can be understood by a wider audience. At some point, a separate formal paper will be written.

Yes, I have worked on a scientific paper with a professor from Berkeley University & presented the same in an International Conference a while ago. Yes, I know how to write a paper. Yes, this paper is deliberately written this way.

The results of this paper are easy to understand for a student of physics, however the path to these results (even though they do not directly influence the result, because the result is a discovery not a proof) will not be agreeable to many who are steeped in current math theory, and unfamiliar with concepts of physics.

Please remember this is a new theory which is similar to; but very different from the current theory. What you know from current theory, will not apply to this new theory where all the concepts are very different. I believe that some amount of "backwards compatibility" is possible with old math, by considering it to be an isolated, single frame math which ignores relativistic limits by defining division by zero as undefined.

The concepts of zero, infinity and "any" are all very different from that of current math, and hence new symbols are used in many cases. Please note that the symbols are fully elucidated within the Unified Relativistic Numeral System section of this paper.

There are many references to relativity and physics all over the paper. Just so the reader knows, these were added later, because nobody could give me proper feedback after understanding this paper in its entirety, so I had to try and find weaknesses and fix them wherever I could find them. I did this, by going back and researching relativity in more detail and reconciling and strengthening concepts and ideas in this paper.

I will be providing the results in the beginning, and then go back so the reader will better be able to understand why the path to these results make sense (aka how wrong we were all this time).

The greatest mathematicians like Ramanujan and Fermat also sometimes provide results without the proof for it. It is interesting to note that I am being forced to do so as well, because I have found there is no other way to get people to understand the paper before commenting on it.

This is how it all started:

Ref: "Division by zero", from: https://bit.ly/3ef0WuS

This paper is dedicated to the legendary Indian mathematician Srinivasa Ramanujan who has inspired Indian students through the years to think out of the box.

"Just because you do not understand something, does not mean that it is untrue."

Reviewing Scientific Papers 101: A Reminder to the new Generation

Comments and critiques are welcome & encouraged so that I can improve or try to resolve any of the issues which arise out of the theories mentioned in this article. In fact, I am distributing my paper as widely as possible to get all commentary around it.

I do have a few pointers for the readers though, because it seems that pretty much everyone has forgotten the scientific method of respectful debates and discussions:

- First read the paper fully.
- Then try to understand it, and think about it a little.
- Clarify with me, what you do not understand.
    - I promise, I will get back very quickly to you, if you ask serious, specific questions respectfully.
    - Then once you are sure "you get it", critique.
- Note that this paper is written in lucid form because it is still a work in progress.
    - I may change contents significantly over time based on commentary.
    - Sometimes, it takes a while to understand the effects of one statement or theory on another.
- Critique specific items so they can be improved or removed altogether.
- "This paper is wrong" – is not useful, or logical.
    - The sentence "… " on page 25 is wrong, is useful and logical.
    - The equation "…" on page 11, is wrong because of X, Y or Z is useful or logical.
    - How do field axioms fit into this theory? is a valid question.
- Yes, I like strong opposing statements which are specific & actionable.
    - I invite them, so I can strengthen this paper.

A lot of the commentary I get is equivalent to:

"I read this paper, I have no idea what you are talking about, I don't understand what X or Y means, so I am going to assume it means Z, and so the entire paper is wrong".

In such scenarios, I suggest you first ask for clarification what is X or Y – you will see that I very quickly update the paper to explain what X or Y is. Then you ask me well, then does this mean Z? Then, I will look into it, and again update the paper to clarify that.

A lot of commentary I get really has to do more with language and people just not understand what is written, or some concepts not defined, rather than actual mistakes which cause me to rethink my concepts or abandon this paper altogether.

Honestly, I would like to get commentary which seriously challenges what I have written here. So, far I have seen none like that. In fact, I have had to extrapolate meanings from statements, and then figure out that this is probably what a person did not understand, and fix the language.

Conclusions

I would like to clarify that I think of the conclusions of this paper as discoveries and not inventions. I believe that almost everything I mention in this section already exist in the physical world of our universe. I just discovered it, and I had a lot of help and inspiration from the giants who came before me including Einstein & Stephen Hawking most major physicists of the 20th century.

The Unified Relativistic Number Theory: Why any number is literally the result of $\circledast \times \propto$

I admit, by the time I was writing this section, I thought this paper was complete. However, as I was going over the different reasons why division by zero is undefined in current math, my resolve only hardened to make sense of the fact that when we multiple indeterminate zero by indeterminate infinity, the result is any number:

$$1 = 2 = 3 = 4 = 5 = 6.. = \circledast \times \propto$$

$$\lim_{\leftarrow \forall \rightarrow} \forall = \lim_{\forall \rightarrow 0} \forall \times \lim_{\forall \rightarrow \infty} \forall$$

$$\forall = \circledast \times \propto$$

I knew, I had to explain this "meaningless", "nonsense" equation to really make headway with everything I had written till this point.

So, I went back to the basics. Most of the insights in this paper came out due to my strong conviction that even if we do not understand an equation, which resulted out of scientific method, and it feels like nonsense to us, it is inherently correct. We do not understand it yet, but it does have some meaning. It is just that we cannot understand it currently.

So, what is the meaning of saying that every number is equal? or that every number is the result of multiplying 0 with infinity?

This equation and sentence are not meaningless, it actually has profound meaning. Everything we have inferred till this point indicates that we cannot ever truly reach zero or infinity no matter how far, we go down the number line in either direction (See: Why zero is also an indeterminate concept).

So, the number line (in the real world) extends in both directions from zero (nothing) to infinity (the largest number ever). Note that we have already shown how zero is actually an indeterminate very similar to infinity.

The theory states that what we define as 1, 2, 3... or our human definition of number is by itself a matter of *scale* (our perception of small, normal and very, very large numbers). So, the values of 1, 2, 3, etc as we perceive it depends upon our relative frame of reference (as the observer) in our universe.

This means that something very small like 0.001 maybe 1 for an ant, and 1,000 maybe 1 for a blue whale.

Another way to look at it is that for human "one" piece of bread is very different than a piece of bread for an ant, a dolphin, a blue whale and an elephant. Sizes are relative, quantities are relative, and a number system which each of these species were to come up with, would sit at different places in the number line as shown in the below diagram. Note that this would be true even if we don't consider the existence of a multiverse. Even in our own universe, within our own planet, this would be true. Hence, the concept of zero, or infinity would have different values for different species:

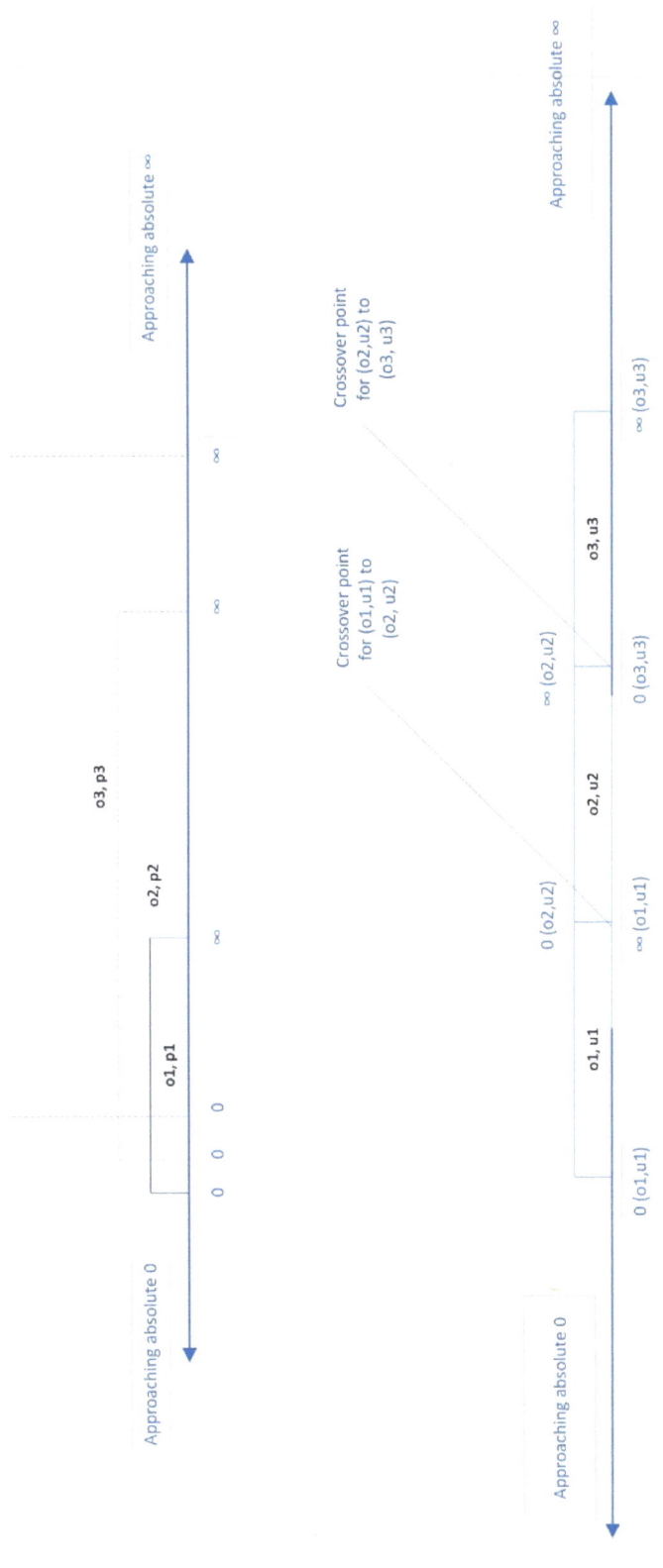

Frames p1, p2 and p3 could overlap each other, or they could be adjacent to each other as well, in which case, the atomic ∞ *for one frame of reference would be where ⊙ began for the other*.

Recently, I have spent a lot of time thinking about this fundamental precept in this paper. Why do observers spring into existence when we have equations approaching relativistic limits like infinity & zero? Even if you do not agree with the statements I have made here, what remains clear is that the number line *does* extend to infinity in one direction and "absolute zero" in the other.

Clearly, we can all agree that neither of these concepts are "reachable" by humans in the physical world (I explain this w.r.t zero later in this paper). If this is true, it also means that because humans defined the numbers 1, 2.. 9 and placed them arbitrarily based on their *frame of reference* or *relative scale* as per where they think they are with respect to zero and infinity; clearly there can exist different numeric systems defined by different species in our universe (whether they exist on our planet or others).

So, from a very logical & scientific principle, even if intelligent life did not exist in the universe, the numerals in any mathematical system would be relative to each other on the number line based on how intelligent life evolved on different planets.

This is why observers come into existence when equations contain infinity and zero, and they produce results which make no sense within a single frame of reference, but make perfect sense when considered across frames of reference and different observers, precisely because the very act of adding infinity or zero to an equation and how it is used in that equation, causes the numbers to naturally move across different frames of reference.

We don't understand the results precisely because for whatever reason, we have decided that the number system we defined is "absolute" which makes no sense when there is no reason why we should be special enough to make our math "absolute" – when we clearly have limits which we cannot approach like infinity or zero. If we were special, and our number system was absolute, then both infinity, zero and beyond would be reachable to us.

This is how we arrive at the below equations:

$$\odot (o1, p1) = \infty (o2, p2)$$

$$\circledast (o1, p1) = \propto (o2, p2)$$

Zero for observer o1 in p1 frame of reference is equivalent to Infinity for observer o2 in p2 frame of reference where p1 frame exists nearer to infinity than p2 frame on the number line

Some people have mentioned that I just use words for "effect" – like "scale" or "scaling". Here is an explanation on what "scale" means (from Wikipedia):

"The subatomic scale is the domain of physical size that encompasses objects smaller than an atom. It is the scale at which the atomic constituents, such as the nucleus containing protons and neutrons, and the electrons, which orbit in spherical or elliptical paths around the nucleus, become apparent..."

"Astronomical scale the opposite end of the spectrum…"

Ref: "Subatomic Scale", from: https://bit.ly/2UgKzFT

I found this interesting reference online while doing my research for this paper:

"*Scale relativity is a proposal for a theory of physics based on fractal space-time. Introduced by Laurent Nottale, **it claims to extend the concept of relativity to physical scales** (of time, **length**, energy, or momentum).*"

Ref: "Laurent Nottale", from: https://bit.ly/2U8zrfy

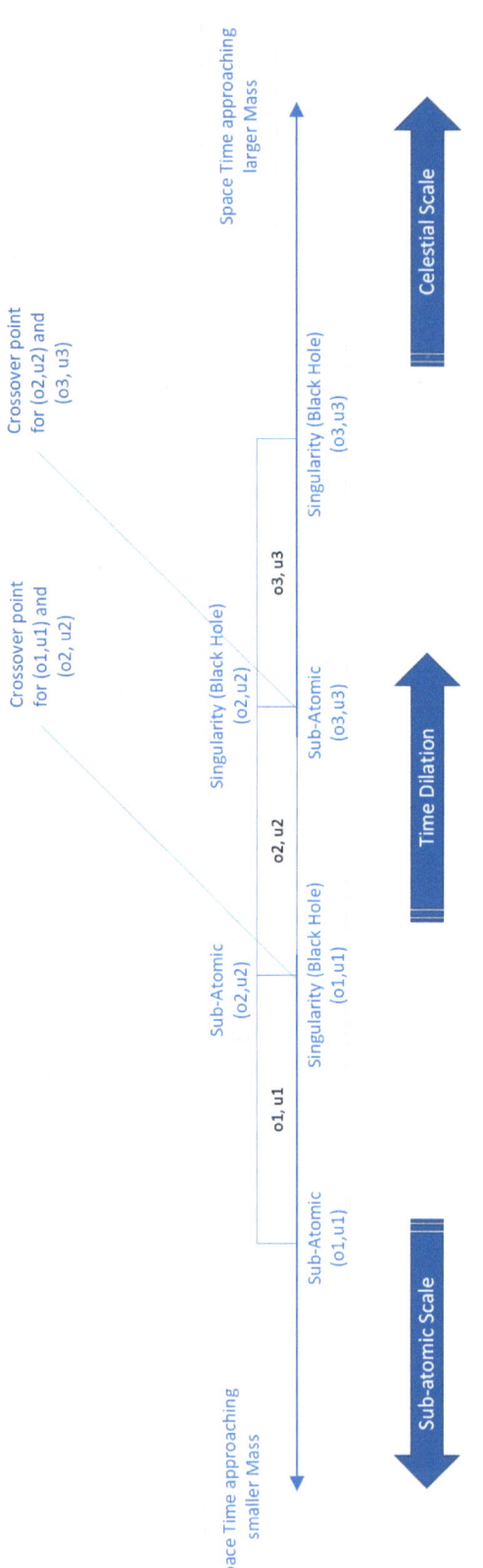

This diagram shows most of the concepts described in the Physics section of this paper. Each observer is in a universe of her/ his own. This is a diagram of the multiverse as it would look if we took the perspective of this paper.

It is the observer's perspective of very small, normal and very large numbers. These equations are *actually* "observer agnostic", because current math does not consider the concept of an observer looking at an equation from different frames of reference – i.e., from the perspective of multiple observers looking at the same equation from different frames all along this number line (not a single observer looking at the equation from a single "frame of reference"), this is why they seem like nonsensical results.

Some people will have a very hard time believing the above statements. For those people, let me clarify that in 1905, the theory of special relativity forever changed our world by stating that **there is no preferred frame of reference. Everything, including time is relative.**

Ref: "What if you traveled faster than the speed of light?", https://bit.ly/2UnFeMV

If you look at the numbers from a relativistic perspective, any number can be constructed based on the frame of reference (sub-atomic to celestial) of some other observer in the universe, because for different observers existing at different frames of reference, any number can be small enough to be considered to be practically equivalent to ⊙, and any other number relative to this observer's "⊙" number at her/ his scale can be considered to be large enough to be practically equivalent to ∞ for another observer.

$$1 = 2 = 3 = 4 = 5 = 6.. = ⊛ \times \propto$$

$$\vee = ⊛ \times \propto$$

$$\vee (o1, p1) = ⊛ (o2, p2) \times \propto (o2, p2)$$

$$\forall (o1, p1) = ⊙ (o2, p2) \times \infty (o2, p2)$$

Whether you look at the equation in its atomic form, or in its indeterminate form, any number for observer o1, in p1 frame of reference can be the result of multiplying atomic or indeterminate zero for observer o2 in p2 frame of referece and atomic or indeterminate infinity for observer o2 in the p2 frame of reference.

Hence, any number can be constructed for an observer o1 existing in frame p1 when you multiply a small enough number from another observer o2's perspective who exists in frame p2, with a big enough number from observer o1's perspective.

Here is a small primer on frame of reference (as per Wikipedia):

"In Einsteinian relativity, reference frames are used to specify the relationship between a moving observer and the phenomenon or phenomena under observation. In this context, the phrase often becomes "observational frame of reference" (or "observational reference frame"), which implies that the observer is at rest in the frame, although not necessarily located at its origin."

Ref: "Frame of reference", from: https://bit.ly/2WtAp7z

Just to note, in our case, we are referring to the observational frame of reference.

This also means that the atomic values of zero and infinity can be approximated for all practical, useable purposes for an observer at a certain frame of reference, as a small enough number which may not be the absolute zero, but adding it to any number makes no mathematical difference at all, and the atomic value of infinity can be approximated for all practical, useable purposes as a large enough number at the same frame of reference for the same observer, that it completely takes over the number being added to it (adding the number to infinity has no mathematical effect on it).

The meaning of every number is equal to every other number is that your 1 is equivalent to a 10,000 from another observer's perspective (who is on a different frame of reference, with larger numbers when compared to your numbers), while your 1 maybe equivalent to 0.0001 from yet another observer's perspective (who is on yet another frame, with smaller numbers when compared to your numbers).

Substitute 1 with 2, or with 3 and extrapolate it, you will see that any number could be shown as equal to any other number from the perspective of different observers. The proof that this is true is the reason why when we chart out an increasing number multiplied by a decreasing number; it always coagulates around specific numbers. It is all a matter of relativity!

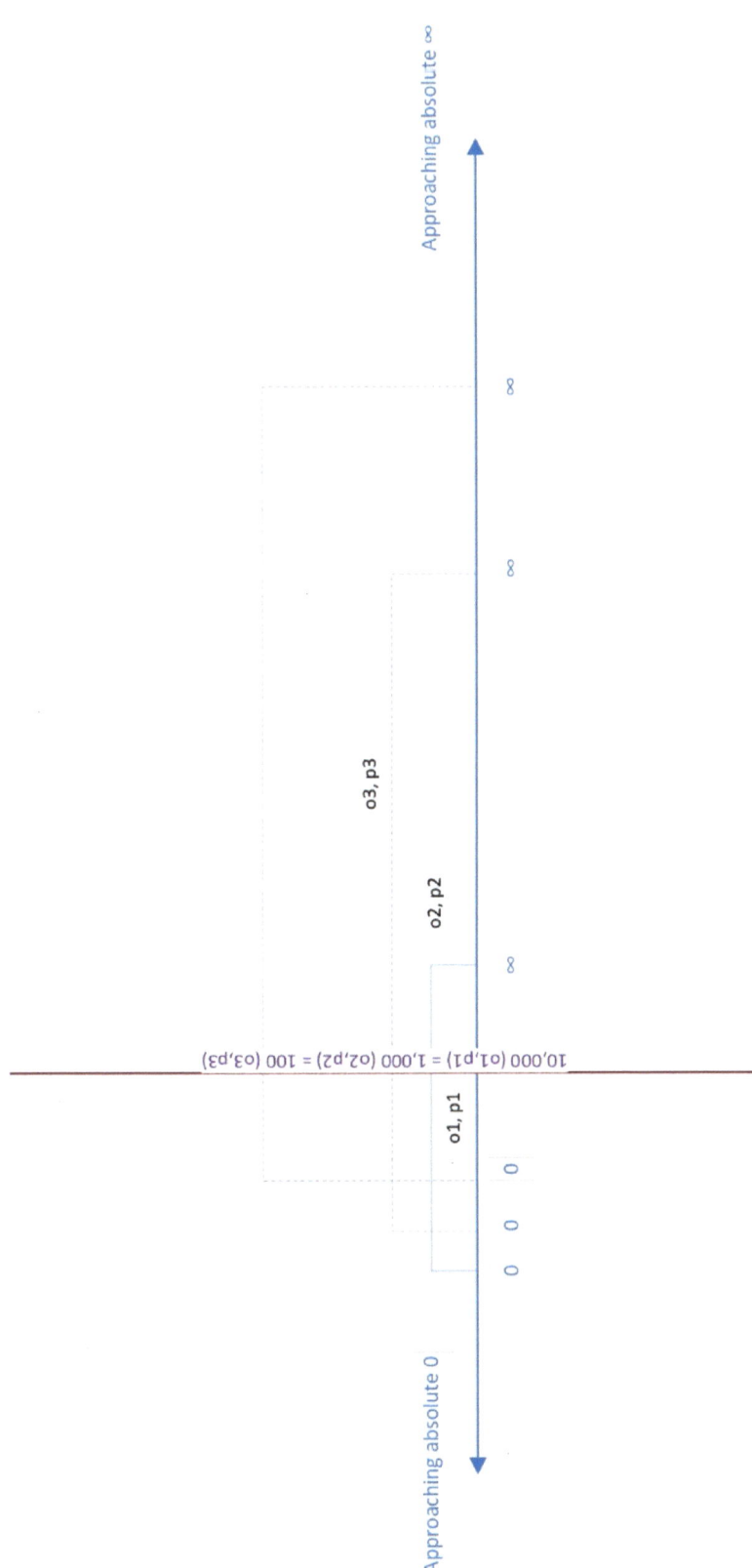

The principle of Number Equivalence for different observers in different frames shown as a diagram.

As an equation, this can be expressed as:

$$1 = 2 = 3 = 4 = 5 = 6 \ldots = \forall$$

$$\forall (o1, p1) = \forall (o2, p2)$$

$$\vee (o1, p1) = \vee (o2, p2)$$

<small>Whether you look at the equation in its atomic form, or in its indeterminate form, there exists a frame 2 and an observer o2 such that a number in that observer's frame of reference would be equal to a completely different equivalent number in another frame 1, with observer o1</small>

The only problem with the "paradox" equations are that, it is not possible to understand the underlying meaning of these equations without the perspective of observers and frames which current mathematics cannot express.

What is equivalent to zero for one observer, can be equivalent to infinity for another observer in a different frame of reference, and vice-versa. Hence, the concepts of zero and infinity are equal to each other, and only seem to be different concepts for an observer who exists within a specific frame of reference. For different observers, in adjacent frames of reference, zero and infinity are practically interchangeable concepts.

I found this interesting excerpt about special relativity while researching for this paper:

*"It has, for example, **replaced the conventional notion of an absolute universal time with the notion of a time that is dependent on reference frame and spatial position**."*

Ref: "Special relativity", from: https://bit.ly/2WyWrFY

The above sentence could be rewritten as follows for this paper:

*"It has, for example, **replaced the conventional notion of an absolute zero & infinity with the notion of zero & infinity that is dependent on reference frame**."*

Naming the Theory

I have named it the "*unified*" number theory because it unifies the concepts of zero, infinity, real numbers & limits along with the basic concepts of Einstein's Relativity bring math closer to the physical reality of the existing universe which are defined by known and unknown laws of physics. This unification, should allow to make significant progress in math & physics, including quantum mechanics as well as the unified field theory which attempts to unify gravity and space time with quantum mechanics.

It is the unified "*number*" theory, because the numerals in a frame p1 are "unified" by combining zero, infinity and any into the main list of numerals as first-class citizens. The unification is also in the fact that zero and infinity are also recognized as normal numerals like 1-9 but on a different frame p1 than the numerals in the frame p1.

It is the *"relativistic"* number theory of, because the numerals are susceptible to relativistic effects when they approach zero or infinity.

I initially named it as *"numeral"* theory, but because a wider audience maybe confused by that word, I renamed it to *"number"*.

Integrating Observers & Frames into the (new) Numeral System

One more thing which has become very apparent is that the concept of observer and frame of reference has to be integrated into the numeral system, so as to be able to properly understand relativistic equations. This results in our always having to specify the observer and the frame in the atomic or indeterminate forms of zero, infinity and any.

This results in the following:

$$\odot(o1, p1) \text{ vs. } \circledast(o2, p2) \text{ (Atomic vs. Indeterminate zero)}$$

$$\infty(o1, p1) \text{ vs. } \propto(o2, p2) \text{ (Atomic vs. Indeterminate Infinity)}$$

$$\forall(o1, p1) \text{ vs. } \vee(o2, p2) \text{ (Atomic vs. Indeterminate Any)}$$

Because of relativistic effects, all numerals should always be written along with the specification of the observer and the frame of reference when any of the numerals which exhibit dual behavior is included in an equation.

The Observer

An observer in a frame of reference, is someone who perceives the numerals 1-9, based on relative scale of these quantities to herself/himself, as well as their perception of the distance between the real numerals 1-9 from zero and infinity.

Relativistic Limits in Numbers

A number in frame p1 is said to approach relativistic limits when it approaches zero or infinity, because it is getting near the cross over point to the next frame p2.

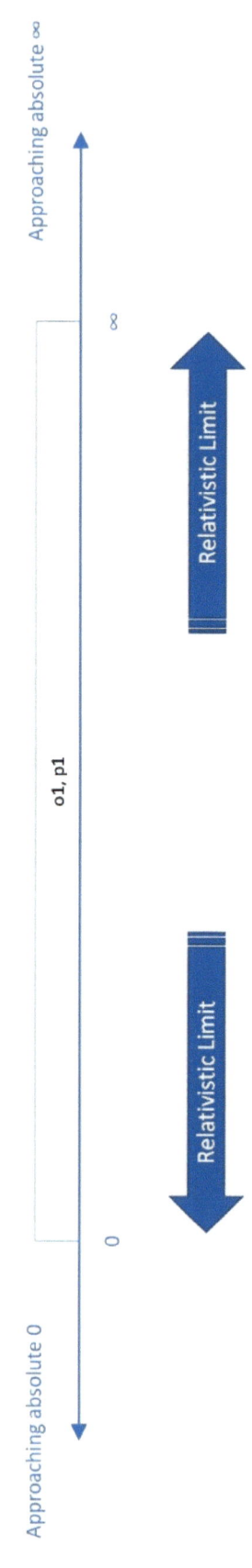

Principle of Relativistic Number Atomicity

Because of the concepts of multiple observers and frames of reference where the cutoff between zero and infinity can be made at any point in the real number line, it follows that zero and infinity ultimately are just human constructs, and from the perspective of the real number line extending to infinity, they are just regular numbers which follow the field axioms in current math within any frame where they fall within atomic zero and infinity.

This can be shown as:

$$\odot(o1, p1) = \forall(o2, p2)$$

$$\infty(o1, p1) = \forall(o2, p2)$$

Atomic Zero and Infinity for observer o1 in frame p1 is equivalent to Any number for observer o2 in frame p2

Zero and Infinity "come to be" as atomic and indeterminate forms only within specific frames of reference where single frame math comes into vogue.

Frame of Reference (Frame)

From a purely mathematical perspective, a frame of reference is a range of numbers which can be placed anywhere on the real number line such than for an observer within that range, the zero "point" is sufficiently small to be considered "nothing" for the observer, and the infinity "point" is sufficiently large to be considered as an unimaginable number from the observer's perspective.

This can be shown as:

$$\odot(o1, p1) + 1 \approx 1$$

$$\infty(o1, p1) + 1 \approx \infty(o1, p1)$$

This concept is very similar to the frame of reference as mentioned in the below article from Wikipedia:

"Far from being simply of theoretical interest, relativistic effects are important practical engineering concerns. Satellite-based measurement needs to take into account relativistic effects, as each satellite is in motion relative to an Earth-bound user and is thus in a different frame of reference under the theory of relativity."

Ref: "Theory of relativity", from: https://bit.ly/2Qvm1b2

I have spent much time thinking about the different frames of references because that is one of the major foundations on top of which this paper is developed. I believe, it is inadequate to explain it in purely mathematical terms only. Let me provide some more concrete examples which will explain exactly why this exists in our universe.

It all boils down to the fact that mathematics per se, is not dependent on the senses of any living organism. Mathematics "exists" by itself. So, whether a living organism has senses like sight, sound, smell, etc. (some of the human senses) or whether it does not (like a virus or a bacteria), wherever it exists, it lives with its own "world view".

There is nothing particularly special about human beings who are one kind of living organism in a universe which has thousands of different kinds of living organisms from all shapes, sizes, and sensory capabilities.

Hence, something "very small" or "very large" as "we define it" is quite different from something defined as "very small" or "very large" for a blue whale at one end of the spectrum and a virus on the other end of the spectrum. In between, there are a wide variety of living organisms so of which can see more than we can, or hear more than we can (like dogs).

So, we really cannot define 0 and Infinity as absolutes from our perspective only.

For a virus, a bacterium or an amoeba, from its frame of reference, its "worldview", zero would be far smaller than what we consider as zero, and Infinity would be far smaller than our conception of Infinity. Hence, mathematical concepts like zero and infinity are by definition relative to the frame of reference of the observer.

Even in current mathematical theory, both zero and infinity are "unreachable". If you go down the number line in either direction, this is why you can never quite reach either zero or infinity. This is just another part of the current theory which neatly fits into our concept of a "frame of reference" because what it is essentially saying is that when you go in either direction, you keep "falling through" different frames of references, and hence you are never able to quite reach absolute zero or infinity because the frames of references never quite "end".

As per this paper, every frame of reference has its own concrete values for zero and infinity. Absolute zero and infinity exist as separate concepts which are best expressed as limits in this paper.

Zero in a frame of reference is a number small enough that it is equivalent to having no effect in the equation within the frame of reference where it is used, and Infinity has the opposite effect of making every other variable in the equation equivalent to having no effect within the frame of reference where it is used.

Approaching Relativistic Limits: Why numbers trans mutate within equations

All particles traveling at speeds near that of light adopt the properties and wavelength of a wave in addition to the properties and momentum of a particle.

Ref: "Light is It a Wave or a Particle?", from: https://bit.ly/38YwyC8

Similar to the above principle: "Any dual natured numeral which is used in an equation where it's value approaches zero or infinity, adopts the properties of an indeterminate number, by trans mutating into a limit approaching the atomic equivalent of that number in the current frame of reference or in alternate frames"

Here is another interesting excerpt about the behavior of light when it trans mutates from a wave into particle form. Such kind of behavior is common in nature:

"When a wave of light is transformed and absorbed as a photon, the energy of the wave instantly collapses to a single location, and this location is where the photon arrives..."

Ref: "Light", from: https://bit.ly/3bfEBvW

I have spent quite a bit of time thinking about why numbers would trans mutate within equations as if they are not "solid", "concrete" entities. How does this make sense?

What humans feel as "intuitive" or "makes sense" are not necessarily the rules of the universe in which we inhabit. If a large mass can "bend" space time to cause the earth to move around the sun, and time dilation bends time itself – the trans mutation of something which humans felt as "concrete", could be a mundane occurrence within this universe.

"A clock that is close to a massive body (and which therefore is at lower gravitational potential) will record less elapsed time than a clock situated further from the said massive body (and which is at a higher gravitational potential)."

Ref: "Time dilation" from: https://bit.ly/2Fmc8tY

If you think about it, what could be more concrete than the concept of time? – the very ticking of the clock, and our experience around the forward motion of time? If time itself is "elastic", pretty much anything else can also exhibit the same behavior around very large or very small concepts.

How can a math which is not built around the same rules as our universe, be able to accurately model it?

We have already shown how equations make sense when numbers trans mutate within equations from atomic to indeterminate forms, while simultaneously crossing over to other frames in some extreme scenarios. This happens in scenarios where equations using dual natured numerals approach numerical relativistic limits. In UNS, this would correspond to the limit of any number approaching zero (indeterminate zero), or the limit of any number approaching infinity (indeterminate infinity).

Approaching Relativistic Limits: Scaling Effects

Considering all that we have discovered till this point, we can infer that whenever an observer $o1$ in frame $p1$ approaches zero or infinity in the same frame, because of the transmutation of numbers, numbers and equations will not behave as expected. This will happen because of the observer effect.

It may be necessary to rewrite equations from the perspective of observer $o2$ in frame $p2$ where the values of zero and infinity move further down or up the number line (corresponding to the sub-atomic world when approaching zero, or the celestial world while approaching infinity from a physics perspective). Such equations will only make sense in it "complete" form where observers and frames come into play.

Relativistic Isolation Theory

I believe that the reason why numbers have the dual nature when they approach zero or infinity is "by design", to ensure isolation between the different frames of reference, each of which may have slightly different, but similar physical laws of nature.

As physical laws change slightly from universe to universe, it becomes essential to isolate each from the other to keep the entire system stable – because the multiverse architecture cannot directly handle a single universe where the laws of physics do not follow consistent rules.

The only way to test out different laws of physics to make a better universe from an evolutionary perspective is to create a set of universes with different laws of physics, and isolate each one from the other, so that no information can "leak" from one to the other. The harder we try to "peek" into and cross over into another universe, the more random things will appear, the "fuzzier" things will become. This could be an inherent behavior "by design" of the universe in which we inhabit which has been postulated to be just one of the many universes in a stable multiverse.

This is necessary because in a multiverse, consisting of different universes with slightly different laws of physics, stability is only achieved in the overall system, if there is complete isolation between these multiverses. This definitely suggests intelligent design, or it could just be the nature of things, just like humans evolved from cells form the primordial "soup" during earths early years.

Translated, different frames of reference have different rules of math, which vary slightly from each other when we move from zero to infinity on the number line. The only way to create a stable math in each frame of reference is to keep each one completely isolated from each other.

Hence, atomic principles which can be "read" without causing any effects because an observer is watching, change into indeterminate waves which are affected merely by the fact that an observer is watching the system. This is also the reason for the existence of the Schrödinger's cat paradox (from Wikipedia):

"Schrödinger's cat is a thought experiment, sometimes described as a paradox, devised by Austrian physicist Erwin Schrödinger in 1935, though the idea originated from Albert Einstein. It illustrates what he saw as the problem of the Copenhagen interpretation of quantum mechanics applied to everyday objects. The scenario presents a hypothetical cat that may be simultaneously both alive and dead, a state known as a quantum superposition, as a result of being linked to a random subatomic event that may or may not occur."

Ref: "Schrödinger's cat", from: https://bit.ly/2Ugzdl7

The proof of the Relativistic Isolation Theory is well beyond the scope of this paper. However, I do want to mention here that this can be proved in software engineering, by creating a system which has to follow certain rules to be stable. In such a system, it is possible to "break" the rules, by violating them, if the system is unaware that the rules are being broken, because the state of the program, is saved into different files, and every state which is incongruent with the other state does not know about the existence of other states which violates the rules of the system, and hence the system is stable on operation.

The Relativistic Isolation Theory can actually be proved to a large extent in Software Engineering. As most people will find it hard to wrap their heads around this paper, and will not believe or understand a large part of it, I figured, it would be better for me to explain how this can be proved. Because, if the relativistic isolation theory can be proved in the real world, in a real application which is still probably being used by some major corporations all over the world, then it proves that, there is some real scientific merit to this paper, and it is not all just abstract "talk".

The Microsoft SQL Server Sync Framework is a framework developed by Microsoft Corporation a few years ago, to synchronize SQL Server databases over the network. It has been superseded by inbuilt support within SQL Server for Change Tracking in the past few years, but a lot of legacy software still exists in the world today which uses this framework.

One of the major limitations of this framework, is that tables cannot be shared between different "sync scopes". What this means is that if we need to divide the tables into different groups like "content", "activity", "ecommerce", "history", "search" etc – and a table cannot exist in two different groups at the same time – that "breaks the rules" of the sync framework and will result in an unstable system.

So, a SQL Server Database which is synced using this mechanism can be considered to be a stable "universe" with certain rules/ laws which cannot be broken.

This was a major problem, because one of the largest customers we were courting those days, required tables to be shared between sync scopes for the project to succeed. Nobody had ever done this before; nobody has done it since – because this would violate the "rules" or the "laws" which govern the sync system.

It took me more than a year of living with the problem to find the solution to this conundrum. Today, I refer to the solution as the "Relativistic Isolation Theory", and I was able to develop a stable, functioning system by using this theory to develop a solution which worked for this customer. Needless to say, it worked well and was used (and probably still is used) today.

To solve the software problem, the key insight I had was the multiverse theory and the stability which each universe had, including the fact that each one was ever so slightly different from the other – but stable – and always isolated from each other completely.

Working from that idea (which I came up with during a run), we implemented a software solution where we created a different set of tables which would store the "violating" sync scopes isolated from each other. So, none of the sync scopes knew that the ones which violated them existed elsewhere. Then, whenever we would need to sync one group of sync scopes, we would load them in the SQL Server tables and run the sync. Next time, when we wanted to sync the violating sync scopes, we would load those in the real tables and sync them as well.

The system worked because SQL Server only saw a set of consistent sync scopes at a time. It never saw the mutually violating sync scopes together. Because of this isolation, the overall architecture was stable, even though individual groups of sync scopes did violate the rules when they were all considered together.

Another reason for the success of this architecture, was also because SQL Server stored the metadata for different sync scope groups in isolation from one another (we are not sure about this and can only surmise).

So, in totality, because the different "universes" a.k.a sync scope groups were completely isolated from each other, including their metadata, SQL Server worked fine.

This is a great example of how we can develop a real world application to solve an "impossible" problem using the "abstract" theories mentioned in this paper.

On the matter of negative numbers

Anti-matter and matter are concepts of modern physics, along with dark matter. This paper is silent on the matter of negative numbers, because we would like to complete the matter of zero, infinity and numerals first.

Negative numbers will be addressed in a different paper, or at a later time as part of this paper (if addressing it becomes a necessity) probably as the Special Theory of Relativistic Numbers.

Effects of Observers & Frames on Physics

As I wrote this paper, I realized that the concepts and theories discovered could perhaps extend to physics as well. So, I added this section so that my stream of thought regarding these matters can be completely unwound as possible insights into some major issues in physics.

Although pure physics is not within the scope of this paper, based on the behavior of numbers in the physical universe, it is very much possible that the sub atomic world from the perspective of observer o1 using a math on frame p1, contains the cross over point to the celestial world (of heavenly bodies) for observer o2 using a math on frame p2. Also, the celestial world from the perspective of observer o1 in frame p1 would thereby contain the cross over point to the sub-atomic world of observer o3 in frame p3. Perhaps Black Holes are that cross over point. That should probably be the object of study for another paper.

Time Dilation in adjacent frames

Clocks near a mass or other source of gravity run more slowly than clocks which are farther away.

Ref: "time dilation", from: https://bit.ly/33qUAEH

If the frames are organized in such a fashion, adjacent to each other, where observer o2 in frame p2 refers to us:

$$(o1, p1), (o2, p2), (o3, p3)$$

Based on the general theory of relativity in physics, this would imply that time moves faster in frame p1, as compared to frame p2 (because p1 is beyond the sub-atomic world for frame p1), and time would move slower in frame p3 as compared to frame p2 (because p3 is beyond the celestial body world for frame p2).

Black Holes as Crossover Points to adjacent frame

If this is true, that would mean that time slows down near black holes. And this is true as expected, because the flow of time slows down when we get closed to a black hole.

Ref: "What are Black Holes?", from: https://bit.ly/395zGff

If the crossover point for frame p2 into frame p3, enters into the sub-atomic world for frame p3, then, it is expected that any matter entering a black hole and passing through it, is broken down into sub-atomic particles when it reaches the other side.

Ref: "The strange fate of a person falling into a black hole", from: https://bbc.in/2Wovu7V

The behavior of Light in a Black Hole & The Relativistic Isolation Theory

Because of the concept of Relativistic Isolation mentioned earlier, it would also make sense that even light cannot escape from a Black Hole (thereby ensuring complete isolation between multiverses).

Ref: "Black Hole", from: https://bit.ly/3d7mKZT

Multi-frame Theory & Black Hole Singularities

As per the theory presented in this paper, the singularity, would be the cross over point from frame p2 into the frame p3. This is very similar to the theory in the referenced article.

Ref: "Black Holes Could Be 'Back Doors' to Another Universe, Say Physicists", https://bit.ly/2UgRb7c

If you find the directional nature of the universes as mentioned in this paper as strange, I found this very interesting article by the University of New South Wales physicist John Webb who is proposing that there could be a directionality to the universe. Which means that instead of the universe expanding outwards in all directions, it could have a dipole structure like the north and south poles of a magnet.

Ref: "The Laws of Physics may break down at the edge of the Universe", from: https://bit.ly/2zB3xjI

The interesting question is then, how come our universe could have multiple black holes? Why not just two – at the left and the right? I believe this is because space is three dimensional, so the "ends" extend from all points in the X, Y & Z directions as well.

Concluding, I can say that current theories of physics do seem to validate the basic assumptions and results of the theory in this paper.

Intuitionist Mathematics

Ref: "What Einstein May Have Gotten Wrong", from: https://bit.ly/2xgkAHj

I read this paper recently about the work of swiss physicist Nicolas Gisin using Intuitionist Math. Here are some of the ideas in this paper which are similar to my thinking not just in this paper, but in general:

- The article mentions that if numbers are finite and limited in their precision, then nature itself is imprecise and hence unpredictable
    - I would change that sentence on the basis of this paper to: "Numbers are finite and limited in their precision based on the senses & relative intelligence of the observing species – hence for any observer existing in a frame, nature itself will seem to be imprecise and hence unpredictable based on their own sensory limitations".
    - *It is amazing to me how physicists always tend to forget that; our experience of the universe will always be limited by our senses.*
    - There is an inherent beauty in my theory because in some ways it merges biology, physics and math together to create a truly unified theory which takes into consideration some of the major sciences in the world today. It is not just pure physics & math.
- It is quite amazing to read the work of L. E. J. Brouwer, who saw math as a construct – where numbers are constructible, their digits calculated, chosen or randomly determined.
    - This paper says the exact same thing. Except, it explains that the reason why this is true is because in different frames of reference, different observers choose to construct math as they see fit, based on the limitations of their senses as well as intelligence.

- I am also dumbfounded to read pretty much exactly the same thoughts as I had while writing this paper that numbers are finite, but also processes. They can become ever more exact as more digits reveal themselves in a choice sequence.
    - This pretty much parallels what I mention here that we have determinate zero and infinity, but we also have the concept of indeterminate zero and infinity. In the latter case, we never quite reach the exact value of zero or infinity, but we can "move" towards those values. I use the word limits, while the article calls it a choice sequence. But, we both refer to pretty much the same idea.
- I found it interesting to read that General Relativity implies that information is destroyed in a Black Hole, while Quantum Mechanics says that it is preserved.
    - This pretty much corresponds to my thoughts on the Unified Field Theory that Black Holes are the boundaries between different universes. The universe on the other side has bodies with significantly larger mass than ours.
    - As mass enters the Black Hole it gets destroyed and beyond the singularity exists the sub-atomic world of the next universe.
    - The entire system, including time dilation has been designed so that no information can pass without being completely destroyed from one universe to the other, because all universes are built on the same template, hence to have a stable multiverse, where each universe has different, but internally consistent physics within themselves, isolation between them has to be built into the overall system by design.
- I also struggle with the "tail problem" as mentioned in this article. It truly is tough to figure out when a number behaves in a determinate vs. indeterminate form. I have resolved that problem by indicating that this is based on the effects of scale (in the case of infinity). For the cases of very small numbers like the example in the article, this paper provides two explanations:
    - From a relativistic perspective, in terms of the human observer, the number would be zero because it is exponentially small. This is similar to saying that "for all practical purposes, from the frame of reference of a human observer, the number is equivalent to zero as zero is defined in our frame of reference".
    - As per this paper, an exponentially small number approaching zero would only start having an effect when it interacts with other dual natured numbers or is used in an equation in such a manner as to start impacting the end results because of the effects of scale.

A Unified Field Theory

As per the theory presented in this paper, a unified field theory would basically show that black holes lead to the sub-atomic world in an adjacent universe (we call that a frame in this paper), where the mass of bodies would be heavier than in our own universe. Also, if we peer deeper into the sub atomic world of our own universe, we would ultimately see the other side of a black hole from another universe, where mass of bodies is much smaller than ours.

Explaining the Complete Lifecycle of a Star

Recently, I have been thinking more about a complete explanation for the entire lifecycle of a star using the theories presented in this paper. I think I have made a breakthrough recently, and this is described below:

- Stars have very large mass as compared to other visible bodies in our universe.
- Some stars are very large, and hence have very large mass.
- As we know, space time "bends" in proportion to the mass of a body.
- This is similar to a ball on a sheet of cloth – the heavier the ball is, the more the cloth gets depressed.

Note that if the surface of the cloth is our universe, then clearly there is something under the "cloth" which corresponds to the next universe in the multiverse.

- We can say space gets depressed & time slows down when the mass of a body increases.
- Beyond a certain mass, space gets depressed so much that it starts to literally curve around the large mass (like a heavy iron ball on a bedsheet).

Note that this also means that the entire space time around the depression resembles a funnel, the bottom of which is the heavy star which is depressing the space time. What we think of as a black hole "sucking" anything into it, is nothing but space time curving into a funnel and things "falling" into the funnel into the star at the bottom of the funnel.

The funnel is what we refer to as the event horizon.

As the diameter of the opening reduces when you go towards the bottom of the funnel, it crushes everything going into it, as it falls deeper into the funnel. When mass falls into the funnel and collects at the bottom, the star at the bottom, gets heavier & heavier – this is why more and more mass collects within the singularity within a black hole.

This shows several things:

1) Why do black holes "suck things up"?
2) How does the event horizon form?
3) Why does more and more mass start to accumulate in a smaller and smaller space as the Black Hole gets "larger" in terms of mass?

It is important to note that in this example, the "cloth" is also the boundary of our universe. When a depression forms in it, that means the "slope" of the "cloth" is also depressing everything within the Universe which lies in between the two "slopes" both flowing towards the depression. Obviously because the Universe spans extremely large distances, any depression would affect bodies near the Black Hole.

Because time moves slowly towards the depression, this also means that any movement towards the event horizon is slow. The diameter of the event horizon should be very large in the beginning, getting smaller and smaller as the mass increases within the "ball".

Why doesn't the entire Universe get sucked up into a Black Hole?
This does not happen because time slows when space "depresses". This is one part of what stabilizes our universe. The second part is because the depression does not slope the entire Universe towards it, because of the extremely large surface area of the Universe.

Time slows, around Black Holes, hence the Universe is stable, thereby we exist.

Note that the Doppler redshift indicates that the Universe is always expanding. This also means that the effect of specific Black Holes is very localized in their areas.

Think of it as a balloon filled with water suspended in space, and rocks collect at various parts of the balloon and it tries to break through but that process takes a very long time to complete because of time dilation. It may eventually break the balloon and the balloon will rip apart and explode – but this is happening so slowly that it will take billions of years to complete within which time entire civilizations can arise and die, many, many times!

What happens Next?

Once the mass within a Black Hole has increased beyond a certain size, because of a combination of increasing mass combined with the force of the depression formed in space time, it rips through the Space Time in our Universe. However, that "hole" is extremely small, and hence it opens up an extremely small hole into the sub-atomic world of another Universe.

The reason why the celestial bodies grow, then get smaller, become Black Holes and then rip into the sub-atomic world of another Universe in the multiverse is because of the theory of relativistic isolation.

Time slowing almost to a halt, prevents the source Universe from collapsing into the hole. And the hole leads to the sub-atomic world in another Universe, with the hole being so small that it cannot really be detected or "peered into" from the other side, because it is so small, it is equivalent to "zero" in that Universe.

The Proof is in the Pudding

For those who think that nothing in this paper makes sense, how do you explain that I've built software systems which have used some of the principles mentioned in this paper? Indeed, it is possible to use some of the ideas mentioned here in the real world.

A few years ago, we had a major requirement from a top retailer in the United States. Nobody has solved this problem before because basically we would have to break the rules of an existing framework to realize the project. Every single person who had tried to solve it had failed including the Chief Architect.

I worked on the problem for two years, living with it, all the time. And one day I figured out the Principle of Relativistic Isolation which I mention in this paper and went to my boss with it.

He understood one part of it. We implemented it successfully - and that is the only system I have ever heard of which works stably while still breaking the core rules of its own underlying framework.

I tried to explain how it works to both my boss and the DBA (a very, very intelligent guy). Nobody could understand it fully. Perhaps because why would a software person ever have had to learn Relativity or Quantum mechanics? I suppose, I knew these subjects more than many other people around me because I have been interested in these concepts since childhood.

It is possible you have used a website whose backend contains my software, because we had and possibly still have thousands of websites used by small, medium and large companies built on our product. It is possible the website of your college or your bank was built using our product, and if they are still using our software, they are still using the code I built on a daily or weekly basis to do some essential backend activity.

It is true that no matter how much I explained, nobody else could understand it. Nobody else could maintain it, nobody else could figure out what made it tick. I am not a DBA, but a year later, when a bug was found (because the DBA never really understood what we did), nobody else (including the DBA who helped develop the system) could figure out how to fix it. Nobody else wanted to touch what was already working fine for many clients. So, I had to figure out where the code was not following the proper isolation principle and fix it – and we never heard of any bugs in the software since then!

I've mentioned this in Job Interviews, as "the hardest problems I have ever solved". And they have asked me for clarification, and I've explained it – and got hired.

This paper is not built on smoke & vapors. The concepts mentioned here can be used to solve practical problems when the situation arises.

Quantum Number Theory/ Rule of Indeterminate Values

All the three scenarios mentioned below can be considered to be "scaling effects". I.e., matters of scale start becoming important, causing the numbers in the equation to transform into their indeterminate forms, which follow different mathematical rules than the real numbers 1,2,3..

Operations done between in-determinate numbers will always result in indeterminate "moving" numbers.

The presence of an in-determinate number in an equation (⊛, ∝, ∀) can transform atomic numbers in the equation to manifest into their in-determinate forms.

Operations in equations can transform atomic numbers into in-determinate numbers by increasing the size of the effect the atomic number has on the other numbers in the equation. 1/0 is a good example of the operation (division), increasing the effect of 0 on the numerator, transforming it into its indeterminate form ⊛.

Hence, even though $1/⊙$ or even $1/∞$ is allowed, the corollary of that which is: $1 = ⊙ \times ∞$. This equation transforms into: $∀ = 0 \times ∞$ because the right-side operation is done between two indeterminate values, and *in such cases the determinate value trans mutates into an indeterminate value.*

For the vast majority of normal cases, we encounter on a daily basis (like buying groceries), ⊙ is atomic:

$$1(o1, p1) + ⊙ (o1, p1) = 1 (o1, p1)$$

Clearly, when an equation contains more than one indeterminate number, then these numbers tend to behave more indeterminate than atomic. This can be proven by graphical analysis, and hence is an observation, rather than a statement of fact:

$$\frac{1\,(o1,p1)}{⊙\,(o1,p1)} = ∝ (o1,p1)$$

$$\frac{1}{\lim_{∀ \to 0} ∀\,(o1,p1)} = \lim_{∀ \to ∞} ∀\,(o1,p1)$$

∝ tends to behave indeterminate in most usages because of scale differences between itself and any other imaginable number. Hence, the behavior similar to:

$$1\,(o1,p1) + ∝ (o1,p1) = ∝ (o1,p1)$$

$$1\,(o1,p1) + \lim_{\forall \to \infty} \forall\,(o1,p1) = \lim_{\forall \to \infty} \forall\,(o1,p1)$$

⊛ tends to behave indeterminate when even though it is small, because of the way it is used in an equation, it has an outsize effect on the outcome (like when because it is so small, and we are asking to divide an orange into pieces that small, we get an infinite orange pieces). The best example of this is:

$$\frac{1\,(o1,p1)}{⊛\,(o1,p1)} = \frac{1}{\lim_{\forall \to 0} \forall\,(o1,p1)} = \lim_{\forall \to \infty} \forall\,(o1,p1) = \propto (o1,p1)$$

Because the dual nature of these numbers manifest based on scale (as compared to other values in the equation) or usage which influences the scale (in the same way), this is very similar to Quantum physics vs. Newtonian physics (hence the name).

It is interesting to note that because of the dual nature of ∞, it shows a unique behavior as shown below. Both of which can be considered to be true:

$$\infty\,(o1,p1) - \infty\,(o1,p1) = \odot\,(o1,p1)$$

Atomically, ∞ is a determinate number, and subtracting anything by itself yields zero as expected.

$$\lim_{\forall \to \infty} \forall\,(o1,p1) - \lim_{\forall \to \infty} \forall\,(o1,p1) = \lim_{\leftarrow \forall \to} \forall\,(o1,p1)$$

$$\propto (o1,p1) - \propto (o1,p1) = \vee\,(o1,p1)$$

In indeterminate notation, the same equation yields a different result.

I believe that this explains why both these equations can be considered to be true, rather than as a paradox. If you extrapolate the above equation, the following can be obtained:

$$\lim_{\forall \to \infty} \forall\,(o1,p1) = \lim_{\leftarrow \forall \to} \forall\,(o1,p1) + \lim_{\forall \to \infty} \forall\,(o1,p1)$$

$$\propto (o1,p1) = \vee\,(o1,p1) + \propto (o1,p1)$$

These equations show that the atomic equivalents of the indeterminate numerals are not equivalent to their indeterminate equivalents:

$$\odot\,(o1,p1) \neq \lim_{\forall \to 0} \forall\,(o1,p1) = ⊛\,(o1,p1)$$

$$\infty\,(o1,p1) \neq \lim_{\forall \to \infty} \forall\,(o1,p1) = \propto (o1,p1)$$

$$\forall\,(o1,p1) \neq \lim_{\leftarrow \forall \to} \forall\,(o1,p1) = \vee\,(o1,p1)$$

In indeterminate form, all the above equations are true. This is part of the reason, the indeterminate numerals should be converted into their indeterminate equivalents when used in equations where they behave more indeterminate than atomic.

In indeterminate form, these numbers, utilize different values when they are used together in the same equation. *This is not a statement of fact, rather it is an observation based on the graphs we have already shown in the previous sections, and how we used that to develop meaningful results to the equations we were analyzing at that time.*

The result is the same whether you use different indeterminate numbers together, or whether you use the same indeterminate number more than once, which is what makes this possible:

$$1\,(o1,p1) + \forall\,(o1,p1) - \forall\,(o1,p1) = \forall\,(o1,p1)$$

$$1\,(o1,p1) + \lim_{\leftarrow \forall \rightarrow} \forall\,(o1,p1) - \lim_{\leftarrow \forall \rightarrow} \forall\,(o1,p1) = \lim_{\leftarrow \forall \rightarrow} \forall\,(o1,p1)$$

Used more than once in the same equation, they do not cancel each other out.

Particle-Wave Number Theory/ Duality Principle of Indeterminate Numbers

This states that concepts like $\odot$, ∞ and $\forall$ have a dual nature. They refer to a specific number (atomicity), but also become indeterminate/ "moving values" $\circledast$, $\propto$, and $\vee$ in other cases.

It is important to note that indeterminate does not mean range of numbers like [a,b] or [a..b], the meaning more closely *resembles* "a approaching b" or limit, except that in this usage, the limit is not a function.

$$\lim_{\forall \to 0} \forall\,(o1,p1) = \circledast\,(o1,p1)$$

Indeterminate expression for 0

$$\lim_{\forall \to \infty} \forall\,(o1,p1) = \propto\,(o1,p1)$$

Indeterminate expression for ∞

$$\lim_{\leftarrow \forall \rightarrow} \forall\,(o1,p1) = \vee\,(o1,p1)$$

Indeterminate expression for $\forall$

The indeterminate expressions are useful when these numbers transform from atomic to indeterminate form, in equations where scale becomes a factor. This is a key concept which we will use over & over again to make sense of what used to be equations which result in paradoxes.

What do we mean by scale? It is similar to what happens when the rules of physics transform when encountering sub-atomic scale vs. large bodies. I.e., there are three different sets of theories to handle:

1. Sub-atomic scale (approaching $\odot$)
    a. It is well known that some of the weak forces only become tangible at sub-atomic scale

b. This is similar to how ⊙, has an outsize effect on the equation: 1/⊙, and transforms it into 1/⊛
2. Natural world we inhabit (1,2,3,4…)
  a. This is where the current math and numeral system work fine.
3. Large bodies like planets (approaching ∞)
  a. It is well known that Gravity is only tangible when the size of the body grows enormous.
  b. This is similar to how ∞ has an outsize effect on most equations where it is used, and transforms the equation from $1 + \infty$ to $1 + \propto$

This is consistent with our physical reality in this universe, because we have other examples of systems (like light) which behave in this dual manner. Also, in quantum theory, physics behaves differently when considering bodies with large mass, vs. bodies with very small mass. The universe which we inhabit itself has this principle, so it is natural that our math system would ideally be consistent with this principle.

This idea is very similar to the behavior of mass in special relativity:

*"The word mass has two meanings in special relativity: **rest mass or invariant mass is an invariant quantity** which is the same for all observers in all reference frames, while **relativistic mass is dependent on the velocity of the observer**. According to the concept of mass–energy equivalence, the rest mass and relativistic mass are equivalent to the rest energy and total energy of the body, respectively."*

Ref: "Mass in special relativity", from: https://bit.ly/2xhOwlC

In this paper, the idea of invariant mass is *similar* to that of atomic numbers, and the idea of relativistic mass is *similar* to that of indeterminate equivalents of these atomic numbers. Note however that in this paper, atomic numbers are same only for observers within the same reference frame, and there is no concept of the velocity of the observer.

Because of the theory of duality, it is therefore necessary to define slightly different characters to differentiate between all these concepts:

- 0 with "dash" above it — for atomic 0
  - Temporary symbol: ⊙
- ∞ with "dash" above it — for atomic ∞
  - Temporary symbol: ∞
- ∀ with "dash" above it — for atomic ∀
  - Temporary symbol: ∀
- ∀ with a left arrow above it pointing to 0 is the indeterminate notation for 0
  - Temporary symbol: ⊛
- ∀ with a right arrow above it pointing to ∞ is the indeterminate notation for ∞
  - Temporary symbol: ∝
- ∀ with a bi-directional arrow above it is the indeterminate notation for ∀
  - Temporary symbol: V

These will supersede all current usages like 0, ∞ and ∀ so it is clear whether we are referring to the atomic form or the indeterminate form of these numerals in any usage.

The meaning of dividing 1 orange by 0

The meaning of dividing one orange by 0 is that you are dividing it into a smaller and smaller portions, and keep going smaller and smaller (on the basis of the particle-wave theory).

Effectively, this will go down to the atomic level and beyond. I would say if we keep dividing the orange into smaller and smaller portions until you reach 0, the result would correctly be ∞ (if you ever reached that point).

Where do the current Math Theories & Numeral System sit within the new System?

This paper elucidates a completely new theory to explain math concepts in the Einsteinian world of 20th century physics. It is to be noted that concepts in this new theory are different from that in the existing system, and hence these concepts have to be disproved from within the bounds of its own concepts.

Concepts from the old, existing theory cannot be used to disprove completely new and different concepts & rules within the new system, even though some concepts may seem familiar.

From the perspective of the new theory, current math is a single frame, bandaged edifice which sits within an isolated, closed-off (by itself) area of the new system. One closed off frame in a multiverse of infinite frames of reference in the direction of Infinity.

The current system is totally unaware of the existence of observers, frames of reference and the relativistic effects within the larger unified numeral system.

There are many theories within the existing system, which may have similar results to the unified relativistic system, but they are so esoteric, that it is beyond the reach of the common man. It cannot and is not taught in schools, and serious damage is done to young minds who effectively learn to switch off their brains and blindly accept artifices which make no logical sense.

The current math has to rely on the crutches of undefined equations, and meaningless results to turn a blind eye to paradoxes which occur within itself. All these crutches have been carefully designed so that a mathematician can continue his work, blissfully unaware of the universe outside her/ his own human frame.

The problem is that this limited math disables the ability of physicists to solve significant problems in a relativistic universe. Hence, it is necessary to migrate to the new system to solve the mysteries of the universe effectively.

Einstein spent the rest of his life trying to invent the Unified Field Theory in physics. It is very much possible, he failed because he had to work within the bounds of the broken old math.

On the matter of the Field Axioms

name	addition	multiplication
associativity	$(a + b) + c = a + (b + c)$	$(a\,b)\,c = a\,(b\,c)$
commutativity	$a + b = b + a$	$a\,b = b\,a$
distributivity	$a\,(b + c) = a\,b + a\,c$	$(a + b)\,c = a\,c + b\,c$
identity	$a + 0 = a = 0 + a$	$a \cdot 1 = a = 1 \cdot a$
inverses	$a + (-a) = 0 = (-a) + a$	$a\,a^{-1} = 1 = a^{-1}\,a$ if $a \neq 0$

Ref: "Field Axioms", from: https://bit.ly/2XrTupw

Current math is single planar math, the URNS is multi-planar math. Single planar math is an isolated system which is stable within itself because it considers that when it nears relativistic limits, the results of equations like division by zero are undefined.

All numerals in multi-planar math do support the field axioms, however dividing by zero is supported because of the cross over effect when multiple observers and frames come into effect.

Atomic numerals are just numbers (even atomic zero and infinity are merely very small and very large numbers – which we can approximate if we want to), so they support the field axioms.

In-determinate numerals are also numbers, if you look at that from the perspective of multiple observers in multiple frames of reference. Hence, even they are just regular numbers from an absolute perspective and support the field axioms.

I found this interesting excerpt while researching this paper:

"*The incompatibility of Newtonian mechanics with Maxwell's equations of electromagnetism and, experimentally, the Michelson-Morley null result (and subsequent similar experiments) demonstrated that the historically hypothesized luminiferous aether did not exist. This led to Einstein's development of **special relativity, which corrects mechanics to handle situations involving all motions and especially those at a speed close to that of light** (known as relativistic velocities). Today, special relativity is proven to be the most accurate model of motion at any speed when gravitational effects are negligible. **Even so, the Newtonian model is still valid as a simple and accurate approximation at low velocities (relative to the speed of light), for example, the everyday motions on Earth**.*"

Ref: "Special relativity", from https://bit.ly/2xdMQd0

This could easily be rewritten with reference to this paper as:

"*The inability of current match to explain the meaning of division by zero demonstrated a fundamental weakness in its numeral system. This led to the development of the **unified relativistic number system, which corrects the numeral system to handle situations involving all frames of reference and especially those equations where numbers approach zero** (known as relativistic* *equations). Today, unified relativistic number system is proven to be the most accurate model of numbers in any frame of reference.* ***Even so, the classical numeral system is still valid as a simple and accurate approximation at human numerical scale (relative to zero or infinity), for example, the everyday arithmetic.*"

"Any sufficiently advanced technology is indistinguishable from magic."

- Arthur C. Clarke

"Imagination is more important than knowledge."

- Albert Einstein

Unified Relativistic Numeral System (URNS)

In the URNS, every numeral, including real numbers has to be specified along with the observer and her/ his frame of reference, *when writing equations involving any of the numerals which exhibit duality*.

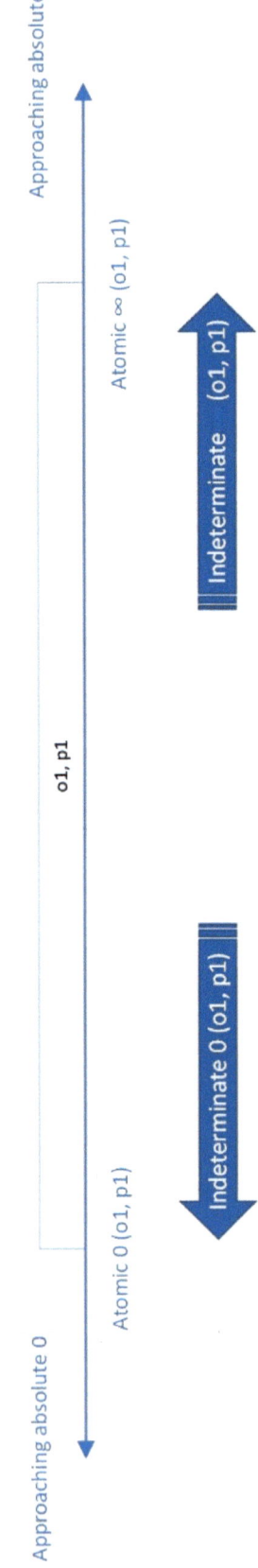

The Positive Real Numbers

In URNS 1, 2, 3, 4, 5, 6, 7, 8, 9… and all subsequent numbers are expressed in mathematical notation, using these digits/ symbols (including zero) in a consistent manner.

These numbers should now be written as: $1\,(o1, p1)$, $2\,(o1, p1)$, $3\,(o1, p1)$… to specify that these numbers are specific to observer o1 in p1 frame of reference, when the frame of reference becomes a factor in the meaning of the equation (when it becomes relevant).

The concept of 0

Humans invented the concept of 0, which is not really a thing, because it indicates "nothing". 0 is a number which quantifies a count or an amount of null size.

In the URNS, 0 can be expressed as an atomic number $\odot\,(o1, p1)$ as well as indeterminate "moving" values $\circledast\,(o1, p1)$ which behave more like a limit than as individual numbers or a range of numbers (Particle-Wave Theory), for observer o1 in p1 frame of reference.

In URNS, 0 is also an indeterminate number just like ∞, but we think of it as a fixed value only because we find it easier to imagine "nothing" than something boundless like ∞.

In URNS, the number line extends infinitely in both directions, across an infinite frame of references at every point in the line. Hence, we can never really reach absolute zero or absolute infinity.

We believe that the original concept of zero in the classical number theory is flawed because humans think they can imagine "nothing". But, in reality, in our physical world, there is really nowhere which is completely empty. We think a closed fist is empty, but there are air molecules there. We can see across a field, but it is not empty. Wherever scientists' thought was "empty" they have found something or the other there – this started with atoms, and we have peered further and further into the fabric of the universe, and we keep seeing something there. It is never fully empty. It is even believed that dark matter exists out there in the universe which has not been detected yet.

So, because nothing is really empty, our concept that we can define "nothing" as zero is flawed, primarily because we have proven in modern physics that things we cannot see or perceive exist beyond the realm of our senses unless we use other technologies to see further like electron microscopes.

In Mathematical form, this would translate to saying that, there is no absolute zero, the way we think of it today. Because absolute zero cannot be reached, as the number line extends infinitely in that direction, across frames of references which we cannot access. This is exactly why the deeper scientist's peer; they see more and more particles within particles.

So, then what should we be defining zero as?

The value of zero in our frame of reference will be equivalent to an extremely small number in our frame of reference. A number so small, that it has no physical effect of it being present in an equation. Which means that for any equation, we can pick a number so small that its value does not matter within its usage – that would be the value of "zero" for our frame of reference.

It is very much possible, that we can ultimately define a single super small value to zero which is so small that it can fit this requirement for any mathematical equation in our frame of reference. That would effectively be the value zero for humans.

Extrapolating, the value of zero in our frame of reference is a number which is so small that it is impossible for humans or any other intelligent organism in our universe to detect when it is added to any other number. It exists at the intersection of mathematics and physics, and some examples of its characteristics would be, a number smaller than:

- Add the amount of decibel to sound, and no organism can detect the decibel added to it.
- Add the amount of transparency to an opaque object, and no organism can detect the amount of transparency added to it.

In addition to characteristics when we can hear, see and feel, the number should be small enough to manifest no behavior which affects our physical universe in any manner. As an example, we cannot see atoms, but atomic structure does affect us in our universe, so the value of zero will be smaller than that of the atom, or the proton or the electrons, so on & so forth.

Zero would be smaller than the size of the smallest object we can observe from the most powerful electron microscope we can build today – or in the future. If we can see it, then zero is smaller than that value. In that sense, we could go infinitely in any direction across any of our senses, but I feel that there are limits to how much we can peer into the universe even with the most sophisticated instruments we have not designed yet.

Using absolute zero as the value of zero, is flawed, because it is beyond our reach, and outside our frame of reference.

Behavior of Atomic Zero $\odot (o1, p1)$

Atomic zero for an observer o1 in frame p1 is a number on the frame p1 which is so small that it has no mathematical effect on any equation. For frame p1, atomic zero is practically equivalent to "nothing". This means that:

$$1 (o1, p1) + \odot (o1, p1) = 1$$

$$1 (o1, p1) - \odot (o1, p1) = 1$$

$$\frac{\odot (o1, p1)}{1} = \odot (o1, p1)$$

Transmutation of ⊙ ($o1, p1$) into Indeterminate Zero ⊛ ($o1, p1$)

Atomic Zero trans mutates into indeterminate zero when effects of scale start to happen, like in the below cases, where it no longer behaves like nothing, because it affects the result of the equation:

$$1 \, (o1, p1) \times ⊛ (o1, p1) = ⊛ (o1, p1)$$

$$\frac{1 \, (o1, p1)}{⊛ (o2, p2)} = \propto (o2, p2)$$

We never write these equations using atomic zero, because that would just be wrong. Also, note how in the second equation, not only did the numbers transmutate, but it zero and infinity also crossed over to another frame.

The concept of ∞

Humans invented the concept of ∞. It represents something that is boundless or endless and something that is larger than any real or natural number.

In the URNS, ∞ can be expressed as an atomic number ∞ ($o1, p1$) as well as indeterminate "moving" values $\propto$ ($o1, p1$) which behave more like a limit than as individual numbers or a range of numbers (Particle-Wave Theory), for observer o1 in p1 frame of reference.

In URNS, ∞ is an indeterminate number, because humans cannot imagine it to be a specific number. It is just the unreachable largest number.

In classical math, ∞ is not a number, but a "concept". However, in URNS:

- ∞ can be considered to be a number, because it shares key characteristics with 0, which is considered to be a number. This is explained in later sections.
- ∞ is a number and not a "concept" because, it appears in many math equations alongside other numbers. A "concept" cannot be fit into an equation with other numerals. Even in Algebra, "x" is eventually resolved to be a number or a series of numbers which satisfy the equation under consideration.
- The single most important reason why ∞ is a number and not a "concept" is because it is the result of existing math equations which are solved by other techniques (like 1/0). A "concept" cannot be the result of an equation where numbers interact with other numbers.
- This paper will prove that $\propto - \propto = V$ (additive inverse of $\propto$).

Similar to the concept of zero, the value of infinity in our frame of reference can also be a defined number in our universe with characteristics which are not too different from that of zero in our frame of reference.

The value of infinity in our frame of reference is a number so large that any number we or any organism which inhabits this universe can imagine, when added to infinity would be equivalent to adding zero to that number – it has no effect on the value of infinity, because any number we can imagine is equivalent to zero by comparison to it.

Again, I believe that this exists in the intersection of mathematics & physics because I believe we can reach a value which exhibits these characteristics. Analogous to zero, any number accessible to us, will have no physical effects which can be read, measured, or be tangible when we add that to infinity in our frame of reference.

Introducing the concept of ∀ (Any)

This paper invents a new concept to add to the existing numeral system: Any Number (∀ being the symbol for it).

Note: The final symbols for all the concepts described in this paper are TBD but will be similar (but simpler) to the ones presently used.

∀ stands for "any determinate number". It is not 0, it is not ∞, it always refers to any number which has a proper, defined, value.

∀ can be expressed as an atomic number ∀ ($o1, p1$) as well as indeterminate "moving" values ∨ ($o1, p1$) which behave more like a limit than as individual numbers or a range of numbers (Particle-Wave Theory), for observer o1 in p1 frame of reference.

The Relativistic Cross-over Principle

Relativistic effects cause the creation of ∀($o1, p1$) in an equation across the boundary of frames from the other numbers in the same equation. This is called the "crossover effect", because a frame cross over happens. Hence other numbers in such equations can always be considered to exist in a different frame of reference (ex: case p2) - ⊙ ($o2, p2$) and ∞ ($o2, p2$) as a rule.

The differences between ∀ as compared to 0 and ∞ are:

- As a limit, the indeterminate ∨ can approach both ⊙ as well as ∞, as opposed to ⊛ which as a limit can only approach 0, and ∝ which as a limit can only approach ∞.
- ∀ refers to a determinate number which humans can pin down like 1 or 2.
- The additive inverse of ∨ is ∨ (Particle-Wave Theory).
- ∀ & ∨ include both positive and negative numbers. This paper does not discuss the existence or need for -∀ or -∨ at this time.

∨ is the limit of ∀ as ∀ approaches 0 or ∞:

$$\lim_{\leftarrow \forall \rightarrow} \forall\,(o1, p1) = \vee\,(o1, p1)$$

The relationship between ∀, 0 and ∞ is shown below. These are referred to as indeterminate form equations and explained in later sections.

$$\lim_{\forall \to 0} \forall\, (o1, p1) = 0\, (o1, p1)$$

$$\lim_{\forall \to \infty} \forall\, (o1, p1) = \infty\, (o1, p1)$$

The limit of ∀ when ∀ approaches 0 is 0. The limit of ∀ when ∀ approaches ∞ is ∞

Field Axioms and Prior Art in Current Math within URNS

None of the concepts in current math come into conflict within the URNS, as mentioned in the section comparing the URNS with current math theory. Please refer to the conclusions section of this paper to get further details about why this is so.

"I did not know quite what to make of it. But the results were so fantastic, that I was quite sure that they could not be wrong, because nobody could possibly have imagined it".

- G.H. Hardy on Srinivasa Ramanujan

The Path to the Conclusions

Some of the sections below do not use the observer – frame usage for the numbers because that was discovered later. Hence, in an attempt to show you how I reached my conclusions, I will retain the original equations as-is, without adding o1 or p1 to them. Note however that, we will have to revisit some of the topics in the upcoming sections, later so that we have the equations in the proper form. Do note however, that many of these equations actually make sense when it is apparent that $\forall$ is in a different frame of reference than the rest of the equation.

Foreword

I accept that there is much within the upcoming sections that a classical mathematician living in a single frame math, will disagree to. I agree that I try to explain things from the conventional math perspective while actually my mind is already thinking in many aspects from a URNS perspective.

This is not really a problem, because I have shown through the conclusions that *classical math is forced to define concepts in such a way that it is stable within its frame of reference*. This also makes it impossible to move forward with any better theory beyond itself, because the rules are bound tight enough by design for isolation.

There is nothing wrong with this, and actually, this is probably even necessary & by design, because single frame math needs to be tightly bound, within rules, so it works within its frame of reference.

So, by design, I have to flout many principles in classical math on the way to *discover* a newer, better math. The path to the discovery of new math, has to be through the old math, through a synthesis of old and new ideas.

This synthesis of old and new ideas, along with the associated struggles and examples shown below may not be perfect, but it does not matter, because this paper discovers an existing relativistic number theory, it does not invent it.

The path to discovery (even if it contains mistakes) is not a proof of the discovery itself, but merely the path which inspired the discovery. It just shows how I thought about existing math, and on the way, discovered a new math.

Introduction

$$\lim_{n \to 0} \frac{1}{n} = \infty$$

The limit of 1/n as n approaches zero is ∞

When we take any number and divide it by smaller & smaller numbers, the result keeps increasing in value, and hence when the divisor approaches 0, the result approaches ∞. This is shown in the below graph:

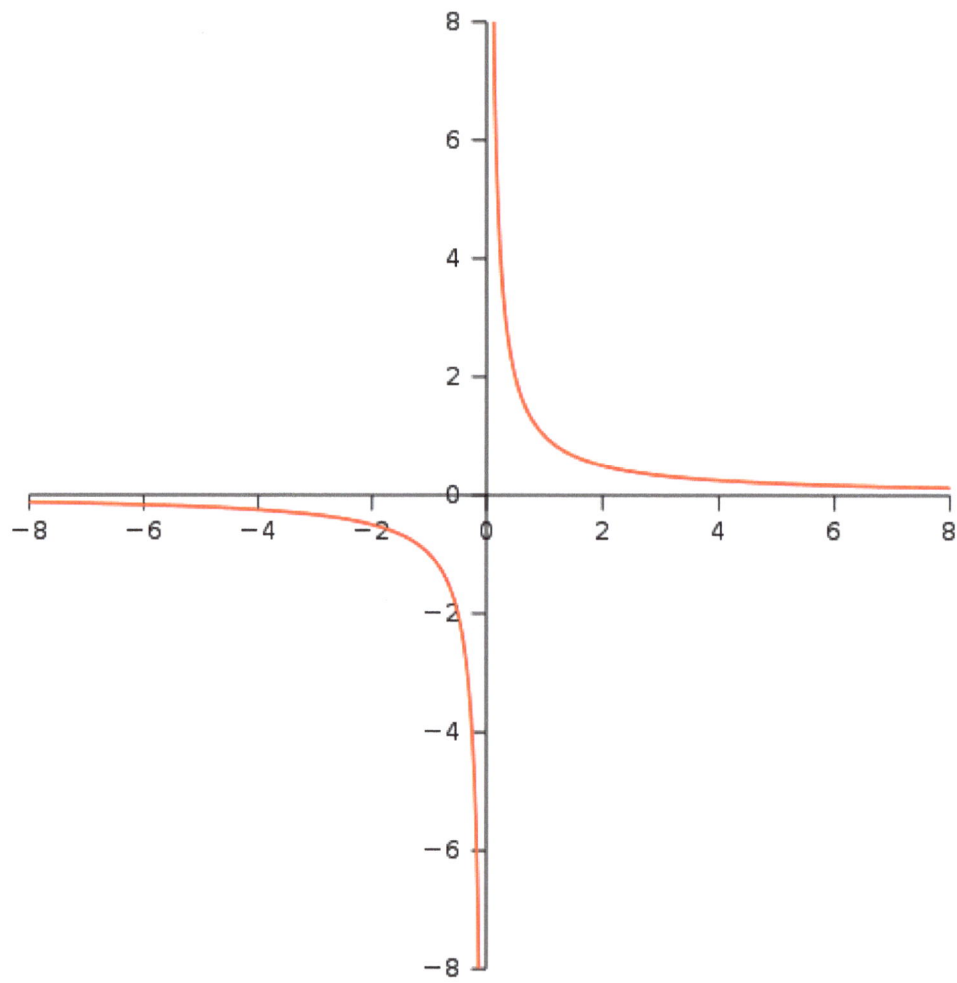

1/0 approaches −∞ as you approach zero from the left, and ∞ as you approach from the right

Ref: "Why is a number divided by zero infinity?", from: https://bit.ly/33vVFe9

When we extrapolate what the graphical analysis shows, the below equation is the only way to express the result using the current numeral system when the denominator is 0 as per the current numeral system:

$$\frac{1}{0} = \infty$$

- Some people say that ∞ is not a numeral, but a "concept".
- This equation which results from well within the theories of current mathematics "as-is" as well as the existing numeral system *without the artificial constructs and bandages which have been put together like glue and toothpicks to prevent the whole system from disintegration.*
- Considered as-is, without bias from prior art, this equation and its result prove that ∞ is a number just like 1 or 0 because it is the logical result of the equation as proved by graphical analysis as well as limits theory.
- $1/0 = \infty$ is an observation of fact, not a premise which "maybe" true, or is "undefined". It cannot be changed by human construct, it just is — whether we understand it or not.
- Another way of saying this is that, it just happens to be a phenomenon we observe in the universe which we inhabit. Whether we understand all the rules underlying the universe — or not.

So, this usage is discouraged **not because, it is incorrect by itself**, but only because it results in a paradox, when we follow the established rules of current mathematics, as well as the existing numeral system, *as-is without the artificial constructs and bandages which have been put together like glue and toothpicks to prevent the whole system from disintegration* as shown below:

$$1 = 0 \times \infty$$
$$2 = 0 \times \infty$$
$$3 = 0 \times \infty$$
$$\ldots$$
$$N = 0 \times \infty$$

This equates to saying that every number is equal to any other number. Because of this paradox, 1/0 is widely accepted to be "not defined", because accepting this as truth, would mean accepting that there are serious flaws within the fundamental concepts of mathematical theory as well as the current numeral system.

This is the first mistake in the currently accepted system.

We are considering that something which can be proven to be true via limits/ graphical analysis as "undefined", **not because it is incorrect by itself, but because corollaries of that result which are fully within the bounds and norms of the established mathematical theory** *as-is without the artificial constructs and bandages* **create a paradox from our perspective (we do not understand the end result, and it seems like nonsense to us)**. Over time, it is apparent that we have built many bandages around this incorrect, flawed system attempting to prevent the apparent complete breakdown of fundamental math, because otherwise it becomes a completely unacceptable system which nobody would agree is plausible.

*This paper argues that **because** the current numeral system provides an apparent nonsensical result (by its own admission nonetheless!!) for the equation 1/0, **it is flawed and incorrect** as it cannot accurately represent the results of an equation which can be proven to be correct with an alternative technique.*

If you read this paper till the end, you will find the reason as to why I call the result "apparent nonsensical result".

The flaw we exposed in the current numeral system is very similar to the breakdown of Newtonian physics when we try to apply it to sub-atomic particles (Quantum theory) as well as bodies with large mass (gravity and space time effects of large masses).

There is in effect no difference between this fatal flaw and the inability (as an example) of Newtonian physics to explain the sub-atomic world or for that matter Gravity. *In both cases, reality fails to fall in line with what the equations show they should be.*

This paper argues that the answer is not to characterize these equations as undefined and retain the existing numeral system, but to invent a new Unified Numeral System which can fix these flaws in the current system.

- The very fact that we designate these equations as undefined signify that the current systems fail at that juncture.

This paper accepts the fact that 1/0 can be proven to be equal to ∞, and moves forward. We will fully articulate the consequences of doing so, interpret said results, make sense of them and explain how these results make sense within the Unified Numeral System. Yes, we will even provide an understanding of the "nonsense" meaning of $1 = 2 = 3 = \ldots = 0 \times \infty$

Is the current Numeral System Infallible?

Some people have critiqued that this paper is incorrect as I am changing some of the most basic assumptions of math. I am not. I am merely taking the current math *as-is without the artificial constructs and bandages which have been put together like glue and toothpicks to prevent the whole system from disintegration*. The road to $1 = 2 = 3\ldots$ leads from the current math theory, and not invented by anyone as an esoteric concept.

Why do I say: "as-is without the artificial constructs and bandages which have been put together like glue and toothpicks to prevent the whole system from disintegration."

I say this because all the concepts of current advanced & esoteric math concepts around zero and infinity were not present when zero and infinity were originally invented (from Wikipedia):

"*Pingala (c. 3rd/2nd century BC[28]), a Sanskrit prosody scholar, used binary numbers in the form of short and long syllables (the latter equal in length to two short syllables), a notation similar to Morse code. Pingala used the Sanskrit word śūnya explicitly to refer to zero*"

"*The earliest recorded idea of infinity may be that of Anaximander (c. 610 – c. 546 BC) a pre-Socratic Greek philosopher. He used the word apeiron, which means "unbounded", "indefinite", and perhaps can be translated as "infinite".*"

Ref: "Infinity", from: https://bit.ly/2Qtpwib , "0", from: https://bit.ly/2w7Dek1

There is a big difference between inventing something new "fully formed" from scratch and taking someone else's invention and adding on things to it because a few people agree to doing it. Consensus is not proof, neither is it science.

Everything in current math ideas about zero and infinity have been added on, later to the original concept. If concepts and ideas can be added to the original concept after the point of inception, anybody (including me) can do the same, if I can develop a stable system which explains everything, and does not result in paradoxes.

The current principles of math are not irrefutable, absolute, undying statements of truth. Math is an invention of man and has many paradoxes which show its imperfections. This paper will make the case that these imperfections in the current numeral system cause results *which seem like paradoxes*:

$$1 = 2 = 3 = .. N.$$

This paper analyzes this paradox from every angle – mathematically, philosophically, logically as well as from a physics perspective trying to understand it, without bias from prior art. At the end of this paper, after exhaustively understanding the new theory, we will articulate a real, profound meaning to the "paradox" because of which current math theory defines 1/0 as undefined.

Analyzing 1/0 = ∞ without bias from Prior Art

Let us get back to the equation again:

$$\frac{1}{0} = \infty$$

- This is proven to be true via limits and verified by the charting technique.
- Note that this equation assumes the definitions given to 1, 0 and ∞ as per the current math theory as well as the current numeral system.
- Logically, it cannot be concluded that this specific equation is "undefined", because it has a perfectly valid result, which can be mathematically verified.

Now, let us look at the corollary of 1/0:

$$1 = 0 \times \infty$$
$$2 = 0 \times \infty$$
$$3 = 0 \times \infty$$

- We cannot explain the meaning of each equation above yet.
- We cannot explain the meaning of all the equations above, even considered as a group (yet).

This further boils down to:

$$1 = 2 = 3 = \ldots N$$

- This result *seems to be clearly wrong*. At this time, it will be good to remember that sufficiently advanced technology seems like magic.
- But we arrived here using the current math theory *as-is without the artificial constructs and bandages* as well as the current numeral system.
- When we move one step above to the earlier step, even those equations by themselves *seem to make no sense*, all we can ascertain logically is that there must be something wrong with our definitions of 0, ∞ and multiplication

Summarizing, because we started with 1/0, the logical conclusion is that there is something wrong with any or all of our definitions for: 1, 0, ∞, division and multiplication — one of these or, all of these.

What does this mean?

Because something we were able to prove, resulted in a completely wrong corollary equation, with a *seemingly* nonsensical result, *some of the currently accepted math theory maybe correct* (because everything else works, as well as the proof of 1/0 = ∞), *however some other part in the currently accepted math theory is definitely wrong* (because we arrived at a nonsense result from the corollary).

What is the wrong with this?

To find the wrong part, we need to isolate the problem. So, we look at how we arrived at the wrong corollary. Where did we start? What were the numbers in the base equation?

They were: 1, 0, ∞, division & multiplication

- Consider 1, which can be said to behave similar to all other determinate numbers. These numbers work perfectly fine everywhere else, and this is a very mundane number, and it makes no difference whether it is 1/0 or 2/0 or 3/0, it all results in ∞, so that cannot be the "wrong part".

- The same argument can be applied to division as well. It is well understood and intuitive. Division is a human construct; it is what we define it to be.
- Multiplication is the opposite of division, so the same argument can be applied to multiplication as well.

That leaves us with 0 & ∞ as the only two possible causes of the paradox.

If you notice division by 0 as well as ∞ both result in the same paradox, which indicate that our definitions for 0 as well as ∞ are either incorrect or incomplete.

The most plausible explanation is that there is something wrong with how 0 and ∞ is defined in the current numeral system (specifically).

A quick note about Wheel Theory here: The Wheel Theory changes the definition of division thereby allowing division by zero. That is too esoteric, because we are essentially changing how a human invented operation works to force-fit the existing flawed numeric system within the bounds of an incorrect assumption ("undefined" equations) to maintain the flawed status quo. This does nothing to resolve the inherent flaws within the current numeral system which forces us to define "undefined" equations to begin with.

Summarizing, 1 = 2 = 3 =… N is a seemingly nonsensical result, and this arises because our definitions of 0 and ∞ are incorrect or incomplete. *This paper argues that the current system is only incomplete, not incorrect.* The Unified *Relativistic* Numeral System is our attempt to complete the incomplete numeral system currently in vogue.

The current Numeral System is the foundation of almost all mathematical theories which came after it. This paper does not aim to create new proofs which bring all the current theories built on the existing foundation under a new umbrella. The scope of this paper is only to fix the fundamental numeral system at hand. Further work will be needed to reconcile all corollaries of the existing theories into the Unified Numeral System.

There are a multitude of numeral systems which exist in the world. **If a new numeral system is better than the existing one, it should be replaced**, just like we replaced the roman numerals with the Hindu — Arabic one.

Ref: "List of Numeral Systems", from: https://bit.ly/2QpLwdG

An Alternate View

I did not introduce the concept of zero into mathematics. Another way of looking at this conundrum is that we can either keep 0 and be forced to keep ∞ as a first class citizen with full rights like 0, or we can forgo 0 completely, and forgo the concept of infinity and division by zero as well.

The flaw in the numeral system started when we introduced 0 as a normal, determinate number and created the concept of division by 0 as a result of it.

We cannot logically invent 0, treat it very similar to any another number and then proceed to say that division by 0 is "undefined". If division by 0 is undefined, then 0 is not a number, it is just a concept which cannot be used in any equation.

The flaw in the current numeral system is that it includes 0, but does not include ∞ as a first class citizen. And to hide the flaws, it defines division by 0 as "undefined", which is similar to eating your cake and having it too.

If 2 - 2 = 0, and hence clearly, nothing is a thing. Then, infinity is also a similar thing. One does not go without the other (because it is the result of 1/0).

If 1/0 cannot be defined even though it is clearly proven to be ∞, then either the concepts of 1, 0, infinity or division & multiplication is flawed. We cannot have our cake and eat it too!

I'm sure that in the past 200 years, mathematicians have developed all sorts of theories like rings and Group Z, and other artifices, etc to try and hide "undefined" from the common man. I'm sure these theories are perfectly reasonable to a master's student in mathematics. I have no problem with that. I am just inventing a better system, which does not require a master's degree in mathematics to understand why 1/0 is "undefined" because of some perfectly esoteric, but reasonable reason. A good system is simple, elegant and can be taught to kids in school or pre-graduate programs.

Prior Art

Many readers, especially those who are not familiar with the more esoteric areas of mathematics seem to be completely unwilling to even consider some of the ideas in this paper.

So, here is an excerpt from the wheel theory which proposes similar ideas — the key difference being that I independently developed this paper and also diverge significantly in a different direction. I thank the contributor who suggested this theory to me for further reading.

"Edalat and Potts [EP00, Pot98] suggested that two extra 'numbers', ∞ = 1/0 and ⊥ = 0/0, be adjoined to the set of real numbers (thus obtaining what in domain theory is called the 'lifting' of the real projective line) in order to make division always possible. In a seminar, Martin-Löf proposed that one should try to include these 'numbers' already in the construction of the rationals from the integers, by allowing not only non-zero denominators, but arbitrary denominators, thus ending up not with a field, but with a field with two extra elements. Such structures were called 'wheels' (the term inspired by the topological picture ⊙ of the projective line together with an extra point 0/0) by Setzer [Set97], who showed how to modify the construction of fields of fractions from integral domains so that wheels are obtained instead of fields."

This paper diverges from the wheel theory in that the concepts are much simpler to understand and its elegance lies in its inherent simplicity. Moreover, this paper is rooted in the natural universe as we inhabit following certain principles which have been observed and proved in 20th century physics.

New Math inspired from Physics

I borrow heavily from quantum theory and Einsteinian physics in this paper. I found this article online which suggests that this has been done before:

Ref: "How and Why did Newton Develop Such Complicated Mathematics?", from:
https://bit.ly/33uSbc2

Confronting the Paradox with the URNS

So, let us take this new concept and move forward with the equations:

$$\frac{\circledast}{\circledast} = V$$

$$\circledast = V \times \circledast$$

$$\frac{\circledast}{V} = \circledast = \circledast \times V$$

One of the key concepts of V is that if in an equation, any number can be put in a specific position, to get the same result, then we can combine all equations containing the determinate number into one equation, by using V to represent any determinate number as shown below:

$$\frac{V}{\circledast} = \propto$$

$$V = \propto \times \circledast$$

This does not necessarily need to mean that:

$$1 = 2 = 3 = \ldots$$

It just means that wherever $\forall$ is used, we cannot substitute specific numbers for it, and that would be the paradox to avoid. We will delve into the subject of this paradox in much detail later.

Which means that, we cannot expand $\forall$ as in the below situation:

$$\frac{1}{0} = \frac{2}{0} = \frac{3}{0} = \infty$$

$$1 = 2 = 3 \ldots = 0 \times \infty$$

This seemingly results in a paradox because we do not understand what this means (yet). Hold on to your thought for now, because we will cycle back later in the article to show that this "paradox" is not really a paradox, and it actually has profound meaning.

Getting back to the current discussion, what we should say is that this translates into:

$$V = \circledast \times \propto$$

For those who think this is a strange equation, consider the below makes perfect sense, because when any number is divided by an ever-increasing number, will result in a smaller and smaller number, so dividing by $\propto$, would logically result in $\circledast$ as shown below:

$$\frac{V}{\propto} = \circledast$$

$$\frac{V}{\circledast} = \propto$$

$$V = \circledast \times \propto$$

Here are more equations based on the usage of $\forall$:

$$\frac{\propto}{\propto} = V$$

$$\frac{\circledast}{\circledast} = V$$

$$\frac{\circledast}{V} = \circledast$$

$$\frac{V}{V} = V$$

$$\propto = V \times \propto$$
$$\circledast = V \times \circledast$$

And more equations:

$$\circledast + \circledast = \circledast$$
$$\circledast + \propto = \propto$$
$$\propto + \propto = \propto$$
$$\circledast - \circledast = \circledast$$
$$\circledast + V = V$$
$$\propto + V = \propto$$
$$\propto - \propto = V$$

Meaning of V = ⊛ x ∞

The meaning of the equation: V = ⊛ x ∝ is what we will try to understand next. The math that we understand and reason about is an inherent part of the universe we inhabit. We are observers here. We can observe clearly that when we divide 1/0 the result is ∝ based on our charting technique, and similarly, we can observe that 1/∝ is ⊛ based on the same technique.

Hence, if both statements are correct, the result of ⊛ x ∝ is V, we just have to make sense of this equation now. **Just because we may not understand what this means does not mean it is wrong or "indeterminate". It just means, we do not understand it yet.**

Math abounds with examples of equations which looked like random esoteric proofs, but which came to be used later as underpinnings of basic physics. So, I did more analysis around the below equation, trying to make sense of it:

$$V = ⊛ \times ∝$$

What you see here is an example of how we can chart the results of what happens when we consider ⊛ to be any number approaching ⊙ and we consider ∝ to be any number approaching ∞ (indeterminate notation):

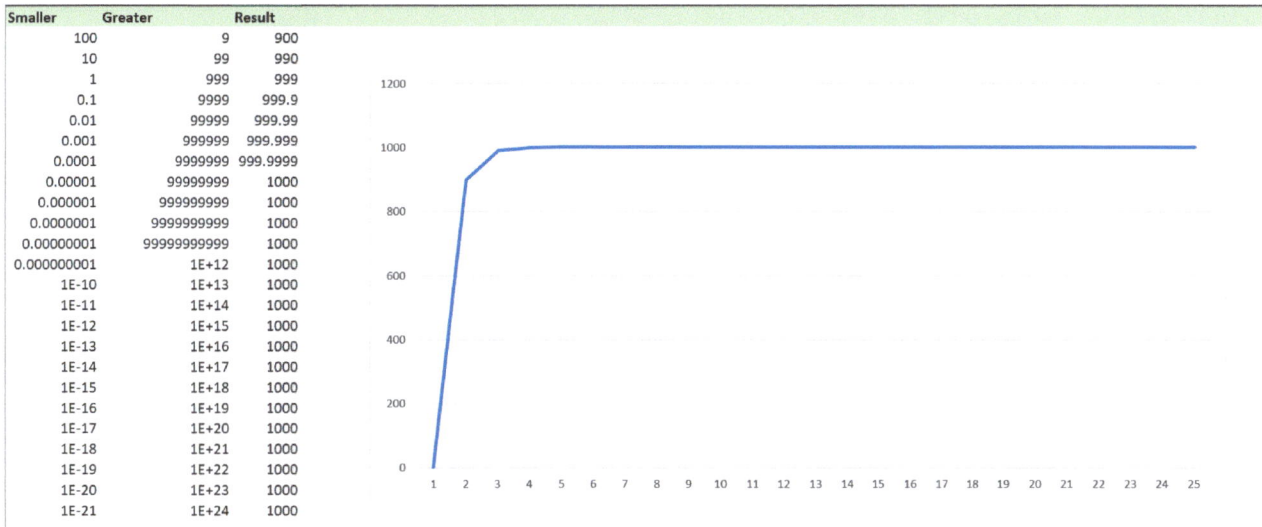

Analysis of multiplication between ⊛ and ∝ (1)

When the numbers go in "opposite" directions, the result always *coalescing* around a single number "in the middle".

However, the point to note is that depending upon where you start, the *coalescing* is around a different number as shown in the next chart.

63

Analysis of multiplication between ⊛ and ∝ (2)

What this essentially means is that the result of ⊛ x ∝ "depends" on where you define the two indeterminate values which we represent as ⊛ and ∝. The result could be atomic number ∀, but the exact value depends on when you give each of these two indeterminate values a specific value. This makes total sense, and is expressed as the equation below:

$$\lim_{\leftarrow \forall \rightarrow} \forall = \lim_{\forall \rightarrow 0} \forall \times \lim_{\forall \rightarrow \infty} \forall$$

$$\forall = \circledast \times \propto$$

Indeterminate expression for multiplying a number approaching ⊙ with a number approaching ∞

What this means is that ⊛ and ∝ can't really be "pinned down" to a single value in scenarios approaching their scale (nearing ⊙ or nearing ∞).

We "pin down" ⊙ only because humans can imagine "nothing", and it is hard to imagine ∞. If we can't determine the value of ∞, then the truth is that we cannot really determine something infinitesimally small either. This is what I will explain in the next section below.

Why is zero in-determinate?

Actually zero, is as in-determinate as ∞ because of this:

0.1
0.01
0.001
0.0001
…
0

We can't quite get there, even though we use 0 to represent nothing.

This directly translates to:

$$\lim_{\forall \to 0} \forall = \circledast$$

The result of trying to understand $\circledast$, directly translates to the limit of any number as the number approaches $\odot$

This is not very different from $\propto$, if you think about it:

999999999
9999999999
99999999999
999999999999
…
∞

$$\lim_{\forall \to \infty} \forall = \propto$$

The result of trying to understand ∞, directly translates to the limit of any number as the number approaches ∞

So, both zero and ∞ can be "pinned" down to one value (atomic notation), but they are also "indeterminate" (indeterminate notation). We do want to mention that the atomic value of these numerals are not equivalent to their indeterminate equivalents as shown below:

$$\odot \neq \lim_{\forall \to 0} \forall = \circledast$$

$$\infty \neq \lim_{\forall \to \infty} \forall = \propto$$

$$\forall \neq \lim_{\leftarrow \forall \to} \forall = \vee$$

The atomic equivalents of these numerals are not equivalent to their indeterminate equivalents

It can be argued that $\propto$ is not indeterminate, and has a distinct value. But, this would be incorrect because by definition ∞ is larger than the largest number we can think of. By definition, it is a moving target = indeterminate.

This link talks about zero also being a concept, just like infinity is a concept:

Ref: "Is Zero Really a Number or Just a Concept?", from: https://bit.ly/2UvCkGf

Summary

So, the real, determinate numbers which can be "pinned down" are:

1, 2, 3, 4, 5, 6, 7, 8, 9…

The indeterminate numerals would be:

⊙, ⊛

∞, ∝

∀, ∨

The reason for the existence of ∨, is because when scale becomes a factor (because of large value associated with ∞, or the equation is such that a small value ⊙ has a large effect), transforming these numerals from atomic to indeterminate form, this transforms the atomic ∀ numeral to its indeterminate form ∨. In the following sections, I will delve further into this topic and provide better insights into this matter, as to why this makes sense.

The question of ⊙/ ⊙

When you try to chart what would happen when you divide ⊙ by ⊙, this is the observation:

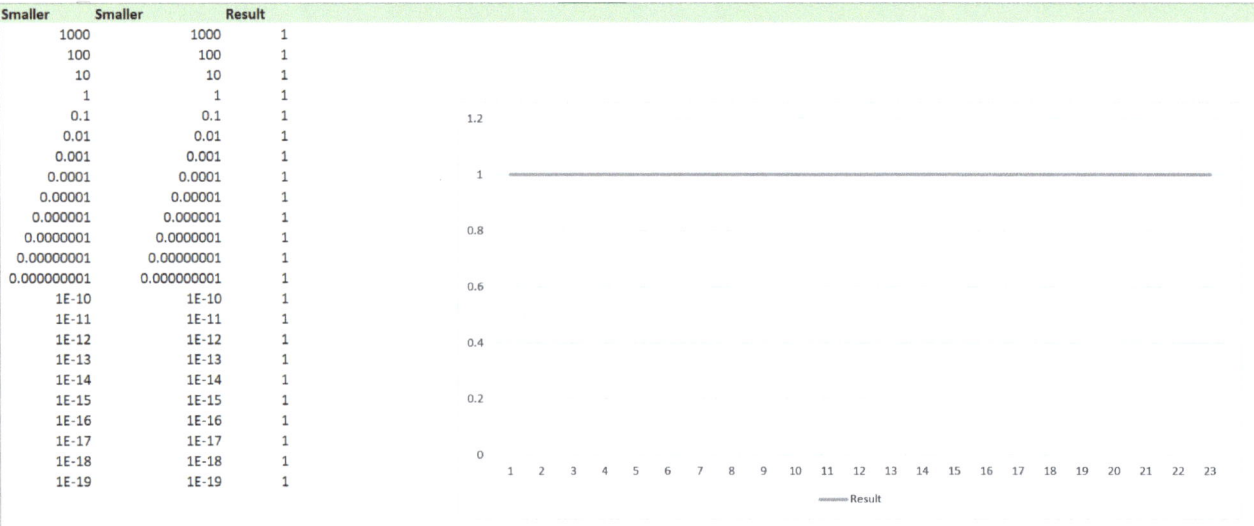

Dividing any number by itself, going down will result in *coalescing* around 1 or a similar number.

So, does this mean:

$$\frac{⊛}{⊛} = 1$$

No, it does not. Because, this graph actually implies that: ∨/ itself = 1. I.e., take any specific number and divide by itself, will give you 1. This chart is the direct expression of the results of one of the examples in which a number approaching zero is divided by another number approaching zero.

$$\frac{⊙}{⊙} = \frac{\lim_{∨→0} ∨}{\lim_{∨→0} ∨}$$

Indeterminate equation corresponding to a number approaching zero divided by another number approaching zero.

66

So, we have to consider further cases of a number approaching ⊙ divided by another number approaching ⊙, by modifying the expression for: ∀/ itself into ∀/ some other number:

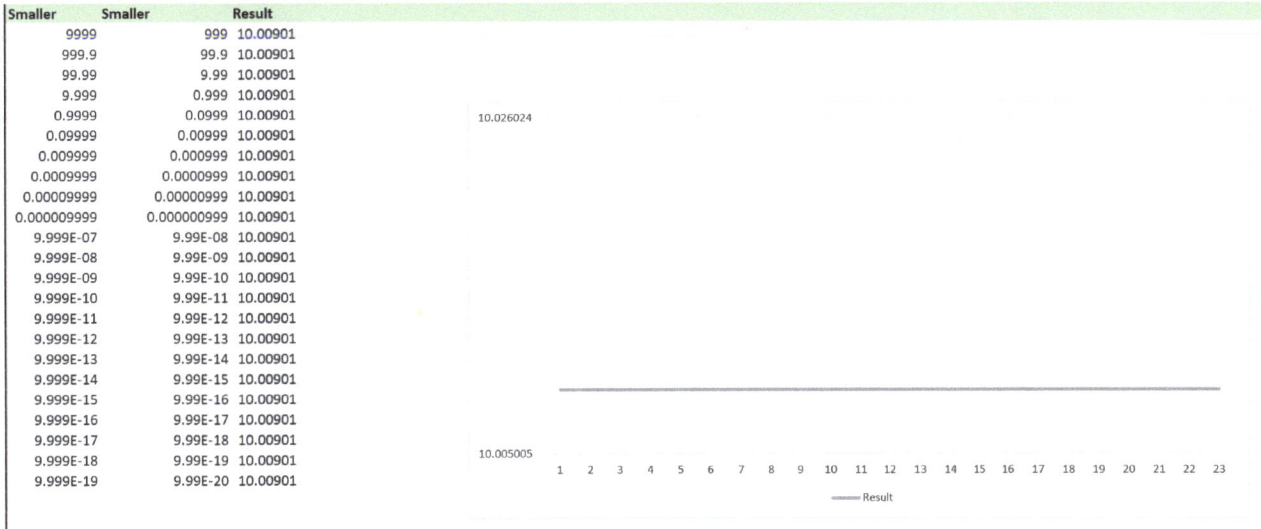

Smaller	Smaller	Result
9999	999	10.00901
999.9	99.9	10.00901
99.99	9.99	10.00901
9.999	0.999	10.00901
0.9999	0.0999	10.00901
0.09999	0.00999	10.00901
0.009999	0.000999	10.00901
0.0009999	0.0000999	10.00901
0.00009999	0.00000999	10.00901
0.000009999	0.000000999	10.00901
9.999E-07	9.99E-08	10.00901
9.999E-08	9.99E-09	10.00901
9.999E-09	9.99E-10	10.00901
9.999E-10	9.99E-11	10.00901
9.999E-11	9.99E-12	10.00901
9.999E-12	9.99E-13	10.00901
9.999E-13	9.99E-14	10.00901
9.999E-14	9.99E-15	10.00901
9.999E-15	9.99E-16	10.00901
9.999E-16	9.99E-17	10.00901
9.999E-17	9.99E-18	10.00901
9.999E-18	9.99E-19	10.00901
9.999E-19	9.99E-20	10.00901

Result of going smaller, with a different set of numbers

Now, we again see *coalescing* around different numbers based on which number you select. Which means, ⊛/⊛ = ∀. Which again makes sense because: ⊛ = ∀ × ⊛, which is where we started, at the beginning of this article.

$$\frac{\circledast}{\circledast} = \frac{\lim\limits_{\forall \to 0} \forall}{\lim\limits_{\forall \to 0} \forall} = \lim\limits_{\leftarrow \forall \to} \forall = \forall$$

The indeterminate equivalent of the charting technique we have described. The Particle-Wave theory is explained later in the paper.

You will see that this is a technique (using the same numbers moving up & down vs. using different numbers moving up & down) I will use many times to explain how things make sense.

Another way to think of why I have used this specific graphical technique, is that when we try to chart 0/0, I am essentially translating that into a graph of two numbers going down towards 0 and checking what the results would be.

When we chart this way, we naturally have to consider scenarios where both numbers are the same vs. being different, to cover the full scope of all scenarios where we are approaching either zero (as in this case), or ∞.

Multi-dimensional Number Theory

This states that the current number system is one dimensional, which results in our wrong interpretation of situations like the below:

$1 = 2 = 3 \ldots = 0 \times \infty$

Because if our number system was multi-dimensional, this could be translated into:

$$1 = 0 \times \infty$$
$$2 = 0 \times \infty$$
$$3 = 0 \times \infty$$
$$4 = 0 \times \infty$$
$$5 = 0 \times \infty$$
$$\ldots$$
$$\forall = 0 \times \infty$$

Our number system cannot properly translate this equation. And I have not shown this figure very accurately either. What I meant to convey is that in these series of equations, substitute 0 and ∞, with a series of numbers going "up" and "down". In that case, the last equation will be the end result of the ones above it.

$$\lim_{\leftarrow \forall \rightarrow} \forall = \lim_{\forall \to 0} \forall \times \lim_{\forall \to \infty} \forall$$

$$V = \circledast \times \propto$$

The indeterminate expression which properly translates the equation: The limit of any number as it moves in either direction is equal to the limit of any number as it approaches zero, multiplied by the limit of any number as it approaches infinity.

Because of the single-dimensional/ atomic nature of the current number system, it does not have a way to represent a multi-dimensional equation, which happens when any of the variables in an equation can have any value like $\circledast$, $\propto$ or V.

0 and ∞ partially allowed us to write a multi-dimensional equation (because they have the dual nature), but without the missing $\forall$, the equation results in a paradox, because we have not yet known of the existence of $\forall$. When we put a specific number like 1 or 2 on the left side of that equation, the equation becomes an incorrect single dimensional representation of a fully multi-dimensional equation (because all values are in fact dual in nature).

This is why, a graphical mechanism has to be used to make sense of and solve these "undefined" equations using the current mathematical tools we have. Graphical representation is the only way which currently allows us to get to the answer of these multi-dimensional equations. **This paper seeks to resolve this anomaly and bring these equations into the unified numeral system without resorting to "undefined" situations, which actually indicate that our math has not yet evolved to make sense of these equations.**

Of course, we have algebra and "x". But we have not extended that into the numeric system yet by introducing another number $\forall$.

Because we don't have ∀, in math, our equations result in what *seems* to be paradoxes like 1 = 2 = 3 = .. etc, because we don't understand this (yet).

One way of understanding that equation is to consider that cases like the above are actually imperfect representations of a multi-dimensional math which cannot be currently represented properly, then this is no longer a paradox.

To be updated further…

References

Wheel theory

Wheels are a type of algebra, in the sense of universal algebra, where division is always defined. In particular…

en.wikipedia.org

Wheel Theory, Division by Zero: https://www2.math.su.se/reports/2001/11/2001-11.pdf

https://en.m.wikipedia.org/wiki/Number_theory

Division by zero

In mathematics, division by zero is division where the divisor (denominator) is zero. Such a division can be formally…

en.wikipedia.org

Infinity

Infinity (often denoted by the symbol or Unicode ∞) represents something that is boundless or endless or else something…

en.wikipedia.org

0

0 (zero) is a number, and the numerical digit used to represent that number in numerals. It fulfills a central role in…

en.wikipedia.org

Numeral system

[A numeral system (or system of numeration) is a writing system for expressing numbers; that is, a mathematical notation...](https://en.wikipedia.org)

en.wikipedia.org

[Limit (mathematics)](https://en.wikipedia.org)

[In mathematics, a limit is the value that a function (or sequence) "approaches" as the input (or index) "approaches"...](https://en.wikipedia.org)

en.wikipedia.org

Notes

Because of so many reviewers unwilling to even consider this new numeral system, we provide a primer on theories:

"A theory is a group of linked ideas intended to explain something. A theory provides a framework for explaining observations. The explanations are based on assumptions. From the assumptions follows a number of possible hypotheses. They can be tested to provide support for, or challenge, the theory."

Here is the crux of the matter as far as this paper is concerned (from Wikipedia):

"Science makes progress by *using one theory* **until it fails**, trying to ***understand why that theory failed***, and *then **making a better theory**.*"

Ref: "Theory", from: https://bit.ly/33vWIe5